The Big Switch

The Big Switch

Australia's electric future

Saul Griffith

Black Inc.

Published by Black Inc.,
an imprint of Schwartz Books Pty Ltd
Wurundjeri Country
22–24 Northumberland Street
Collingwood, VIC 3066, Australia
enquiries@blackincbooks.com
www.blackincbooks.com

9781760643874 (paperback)
9781743822371 (ebook)

 A catalogue record for this
book is available from the
National Library of Australia

Cover and text design by Akiko Chan
Typesetting by Typography Studio

Printed by McPhersons Printing Group.

Thank you, Pamela and Ross, for the gift of a childhood that makes me want to fight for the trees and the bees. Thank you, Selena, for being the sister exploring every tadpole, every turtle, every tree, with me!

Thank you, Arwen, forever my muse.

Thank you, especially, Huxley and Bronte, my half-Australian, half-American kids. You are the generation who will be truly international, for there are no boundaries in addressing this wicked challenge, and tremendous opportunity, called our climate crisis. I fight for you.

And thank you, dear reader, who hopefully will fight like hell with us to fix this mess.

Contents

Preface: 101 million machines

In 2022 the Australian public voted for climate action. Labor and the Coalition received the lowest combined vote since World War II, and six new climate-focused independents were elected to the lower house and one to the upper house. The Greens increased their representation to 12 in the senate and four in the lower house, their highest ever. It was a landslide for climate hope.

I am convinced that in the three years until the next federal election, Australia can embark earnestly on decarbonising the domestic economy (our households and businesses) while putting more money into consumers' pockets. We just need the courage to work out a plan that brings everyone aboard and leaves no community behind.

The reason I wrote this book in late 2021 was to educate voters and candidates about electrification. My friend Mike Cannon-Brookes helped me buy 1000 copies to send to every politician in the country and to leaders across the community, business and public service sectors. My mum joined me on the book tour, clocking 10,000 kilometres in borrowed electric vehicles. We drove Sydney–Adelaide–Melbourne–Hobart–Melbourne–Canberra–Sydney without a hitch, all electrically, which turns out to be very cheap. In every community we visited, the enthusiasm for electrification bowled me over. People are ramping up their emission reduction plans and eager to do more. Teachers and scientists, farmers and politicians are all signing up to the electrification movement. The eagerness of communities, rural and urban, to get on with the job is ... *electric*.

The task ahead will be hard. We need to hit the ground running, with technology that already works, if we are to achieve our appropriately challenging goals. The new parliament has already changed the country. There is a spirit of good faith and cooperation. We now have a federal government that wants to co-operate with the states. We have communities eager to do the implementation.

But here's the first piece of bad news: fossil fuels continue to be subsidised by federal, state and territory governments – in 2021–2022, to the tune of about $11.6 billion a year. A large proportion of this is a fuel excise subsidy, known as the fuel tax credit or FTC, which exempts qualifying individuals and businesses from the typical fuel excise. Directly or indirectly, this subsidises imported oil-based fuels. That means we are subsidising foreign countries, particularly ones engaged in global heating, and – not coincidentally – shady foreign policy. Only weeks before the federal election, the Coalition announced it would help households with the rising costs of fossil fuels caused by the invasion of Ukraine by halving the fuel excise from 44.2 cents per litre to 22.1 cents. Such subsidies are popular in the short term, but they sure aren't sensible long term.

Our annual budget isn't the only place fossil fuels garner government support; there are also tax codes, zoning laws, trade and tariff policies, building codes, road rules, vehicle regulations, superannuation policies, education and training policies, rules that govern union engagement, and even liability laws. Early in the fossil fuel project, in the first half of the 20th century, pulling these levers made sense. We didn't know fossil fuels were going to crash our comfortable climate; we only knew that they were improving our quality of life. Now we know differently, and we need to adjust our priorities.

Incentives and machines

I'm an engineer, so I see the climate problem as a set of machines. A car is a machine. A water heater is a machine. A coal-fired generator is a machine.

A green steel-maker will use a machine. Which machines do we want, and how do we incentivise the Big Switch at every level of society? In the next 20 to 30 years, Australia will replace nearly all of the machines currently in the economy and add some new ones. Bolts corrode. Bearings fail. Dust, rust and dirt have their wicked ways. We keep a few steam trains around for nostalgic Sunday rides, but they don't rule the rails anymore. So it will be with fossil-powered machines. We might keep a Monaro or two running on petrol for posterity, but in 2050, 99% of our machines won't be machines that existed in 2022.

We are going to retire and replace the existing machines anyway, so our policy should be to replace them with zero-emission machines *every single time*. Nearly every time a vehicle is retired, a new machine is purchased to replace it. We should think about each of those purchasing moments and how we help or hinder people in deciding which machine to buy. Is a clean alternative available? Is it competitively priced? Is credit available to those who need it? Is there a workforce to install it? Consider the person about to buy a water heater: which is cheaper, electric or gas? Which is easier to finance? Which has a supporting network of installers? Consider the customer in the car showroom: is the easiest thing to buy electric, or are there economic, cultural and other factors still pushing them towards fossil fuels?

To make this idea concrete, let's name those machines: the 100 million small machines, and the 1 million big machines. It is a useful distinction, because it tells us when, how and by whom decisions are being made. It points to everyone's role in fixing this climate mess.

The small machines: demand side

Step one is acknowledging that there are two sides to the energy and climate debate: the supply side (where we get our energy from) and the demand side (what we choose to do with it in our households and small businesses). Let's look at the demand side first.

There are 10 million households in Australia. There will be 11–12 million by 2050. A few have electric cars, quite a few have electric appliances, but the majority still run on fossil machines. Right now, we have 18.5 million passenger and light commercial vehicles, about 1 million motorcycles, 5 million gas kitchen appliances, 5 million gas water heaters, 5 million gas heaters, 5 million lawnmowers, 5 million chainsaws and whipper snippers and leaf blowers and other small motors, and 5 million barbecues. With some certainty, the majority of those machines will be electric by 2050 because electric machines are simpler, more efficient and better. In total, this makes for 50 million new machines.

Those machines that do the things we enjoy will need support from other machines that generate, store and manage electricity. We'll need about 7 million more rooftop solar systems, 8 million household batteries or their community equivalents, 10 million electric vehicle chargers, and we'll likely need to upgrade 10 million switchboards. That's another 35 million machines, making 85 million machines to electrify the household demand side. There are probably 15 million more in our small businesses and commercial buildings, so let's call it a round 100 million small machines, owned by everyday Australians and small businesses.

Those 100 million machines will be purchased by everyday people in everyday transactions. They might be financed by a car loan, or by a loan against your house, or perhaps paid for on your credit card.

We need to make every one of these purchases easy. This means:

- Having a supply chain for the product: the electric vehicles need to be available.
- Having a well-trained workforce: the tradie needs to know how to install the heat pump or the induction range and be motivated to do it.
- Financing at the point of purchase: the bank needs to be with you when you are replacing that hot-water heater under financial duress.

- Making sure the benefits are passed on to Australian house-holds: we need to align regulations, subsidies and tax codes so that electric machines are as cheap as possible to install, and so that we have a whole-economy solution that is designed to benefit real Australians, not just corporations.

Let's restate the priority bluntly. Electrify our 20 million vehicles. Electrify our 10 million households and replace the fossil machines inside them. Power the whole lot with renewables. Let's learn from our success in capacity building, workforce training and certification, and regulatory cost reduction in rooftop solar energy and apply it to electric vehicles, vehicle-charging infrastructure, home batteries, community batteries, household batteries, electric kitchens, and heat-pump water heating and reverse-cycle A/C or heat-pump space heating. Let's upgrade the distribution grid to support all this. I'm not picking winners. They've already picked themselves. Other technologies might make this story even better, but the basics are already plain to see.

One under-appreciated fact of these demand-side machines is that they are *appreciating assets*. In two senses. They won't lose value as quickly as fossil-fuel machines over the next decade, so are appreciating assets in contrast to fossil machines we might otherwise spend money on. They are also appreciating in their climate performance. Every year that the grid gets greener and our supply chains clean up the performance of these electric machines gets us closer to actual zero emissions. Eventually steel mills running on wind power and battery factories running on sunshine will be manufacturing this future for us. With an emphasis on recycling we can finally close the circle and the clean machines making (and re-making) themselves is our path to sustainability. These are good investments.

The big machines: supply side

The million big machines are the large capital items owned by businesses. These are the purchasing decisions that CEOs and board members labour over: aluminium smelters; blast furnaces; diesel locomotives. Ninety-six black coal mines; three brown coal mines; four aluminium smelters. Nine LNG terminals; 2200 diesel locomotives; three copper smelters; 10,000 mining trucks; 10 coal export terminals. Four oil refineries; 39,000 kilometres of gas transmission lines; 36,000 kilometres of freight rail; 11,000 general aviation aircraft; 2300 commercial aircraft. Twenty oil and gas drilling rigs; 24 coal-fired power stations; 160 gas-fired generators. Twenty-four sugar mills; 22 paper mills; 200,000 light trucks; 370,000 heavy trucks; 110,000 articulated trucks; 100,000 buses. Nickel mines, copper mines, cobalt mines, and the mining equipment to go with them. Add in some other manufacturers, and for argument's sake let's call this one million machines.

Our stock market and superfunds are invested in making the right capital calls for these million machines. We need the technology to be ready, and we need the costs to be reasonable and the risks low. These are the machines that communities worry about, because through them large numbers of jobs are won or lost in a region. For this reason, we need to understand that these big machines, for the most part, are technologies that aren't quite ready to be replaced yet; replacing them needs a plan that considers not only the machine but also its community.

The good news is that these machines last a long time and we have time to plan. The solutions to these challenges are in prototype, in pilot, in proof stage, but not yet in production. We need increased commitment to research and development. We need every young Australian with a good idea to have access to education, training and funding. These aren't software start-ups, in need of short-term investment; we need long-term financing to bring new technology to market. We need to take on more risk, accepting that these projects take time and that some will fail.

This all requires a commitment to retraining and education. Most jobs created by the Big Switch will be for tradespeople and technicians – labourers, not university graduates. We need to be enlisting everyone on the journey.

Climate targets are good, but a climate strategy is better

The newly elected independents are advocating for higher emissions reduction targets. This is good. Australia can and should be aiming higher, somewhere north of 50% and closer to 75% by 2030.

The naysayers say that Australian emissions don't count, that what we do doesn't matter. We produce just over 1% of global emissions, or 4% if you count our fossil exports. But this argument can be turned on its head. We are wealthy relative to most other countries. We are well endowed with renewables. We absolutely will do well in this transition, saving our households money and creating new export industries. We have every reason to go further, faster. (By contrast, if we throw in our lot with the fossil fuel petro-states, our climate pariah status will be cemented.)

To limit global warming to 1.5 degrees Celsius, we need to cut emissions by more than 50% by 2030. But targets get you only so far. Without a strategy and clear priorities, they are empty headlines. I wrote *The Big Switch* to address that, and the book offers three big messages:

1. Electrification powered by renewables is the way to eliminate most energy-based emissions.
2. Our domestic economy is poised to demonstrate this, with the potential to become first in the world for rooftop solar, electric heat-pumps, and electric vehicles.
3. By electrifying industry, we stand to thrive as an exporter of zero-emissions metals and minerals – but the technology isn't ready just yet.

To hit at least a 50% reduction by 2030, the new government needs to encourage these changes. Climate policy needs to shift from vague promises about what might work in the future (hydrogen hubs, green steel, soil carbon) to concrete reductions in the community – in fact, in your household. This is a shift from a 'supply' focus to a 'demand' focus, and it draws an enormous set of new constituents into the conversation. The supply side is just corporations, most of which are still dependent on fossil fuels. The demand side is people, households, small businesses, voters and concerned parents.

To get to zero emissions we need to electrify demand just as quickly as we reduce supply emissions. The shift in focus makes clear that this energy transition will come with the biggest wealth transfer in history, from the traditional suppliers to the traditional consumers of energy. The Big Shift will be great for households and communities if we let it be – and if governments come to the table with policies that help the transition.

Climate action is about community

The 2022 Australian federal election was the climate election. The Greens made big gains. A new band of centrist independents running on climate and integrity platforms – most of them women – upset the Coalition and put the Labor Party on notice. It was also the community election. Tutored by Cathy McGowan and Helen Haines, the original gangsters of Australian community politics, our new independents cut through precisely because they talked to and, more importantly, *listened to* their communities. The communities wanted change. The communities wanted to fix the climate.

By focusing on households in the near-term, the new government would be listening to voters and helping communities. What hadn't dawned on me when I wrote this book was the incredible impact that helping households can have on the communities in which they live. The average Australian household might save $3000 to $5000 a year by totally electrifying their home and vehicle. For the average suburb or small town, those numbers become astonishing. I live in a small suburb north of Wollongong. There are 1000 homes in my suburb, 4500 homes in my postcode. Each year the suburb spends more than $4 million, and the postcode $15 million, on petrol and diesel. This spending creates one or two jobs at the local petrol station, which are really jobs selling tobacco and sugar (three things that kill you in one store!). If we were powering our vehicles with solar electricity, we'd collectively save $3 million in my suburb and $12 million in my postcode every year. That would buy a lot of new classrooms, pay for a bunch of new teachers, and tart up the RSL and surf lifesaving club. New restaurants and stores would open. Tradies would thrive. Some of those savings would end up funding local arts projects. Imagine for a minute your community investing in itself, in local capacity, rather than sending money away to bring in fossil fuels.

For some communities, a shift away from fossil fuels will mean a transition. I live in one. What happens to the blast furnace in Wollongong affects everyone in Wollongong. Will steelmaking and coal-loading simply be shut down, or will we invest in new industries like green steel, off-shore wind production, batteries and photovoltaics? What will happen to the people who dig coal, who move coal by rail, who use coal to make steel? What about communities who use coal-fired electricity to make aluminium? This is why we need to seriously consider the 1 million big machines – the big industries, the export creators, the job makers.

Saying the uncomfortable things out loud

Shortly after the election, the nation's biggest energy market, the NEM, failed. It was a failure of design. The market isn't free. The market isn't up to the job. We need to be interventionist; we can't merely wait for the market to 'do the right thing'. It took the electric vehicle three decades to capture a tiny fraction of global sales, but we need 100% of vehicles and heating systems to be electric as soon as possible. Every intervention that can speed up the market is necessary. As well as providing incentives and rebates to households and businesses, the federal government can dictate the rules of the National Electricity Market. This may be the most important thing the government has to get right.

We can see this in action around the world. US President Joe Biden recently invoked the *Defense Production Act* to mobilise the domestic manufacture of heat pumps, rooftop solar and insulation. The Act enables real action on climate change and will be used to combat inflation and protect America's energy security, while also helping to wean Europe off Putin's gas. Norway has adopted a policy of no new fossil-fuelled vehicles after 2025. The Netherlands has adopted a policy of only electric heating and cooking after 2025. People will still be able to own legacy cars and furnaces in 2027, but they won't be able to replace them with fossil-fuelled machines. Conservative ideologues will say that this is denying people the freedom to choose. Good centrist government denies people the freedom to choose all sorts of things. Australia doesn't let 18-year-olds buy semi-automatic weapons, and that has turned out to be a great idea. Climate change isn't as dramatic as a school shooting, but it is going to harm an awful lot of schoolchildren in the coming century.

If we really want to limit global heating to 1.5 degrees – and that is what we should want – the reduction targets should be 60–75%. This is not an unachievable goal, but getting there will require a holistic reform of our energy system involving all levels of government. We need a whole-government response in which Australian citizens are the beneficiaries of the windfall that climate action will bring. The Australian household hasn't had a very powerful seat at the negotiating table in the Australian energy market so far. That must change.

Introduction: Why this book?

'We're not alone. Good people will fight if we lead them.'
– Poe Dameron, Star Wars

The kids have it right. It's a climate emergency. It is not a drill. Most scientists view 1.5°C of global warming from pre-industrial levels as the temperature we should not exceed if we wish to keep the world's biological and ecological systems functioning largely as they have during the flourishing of our species over the past 10,000 years. 1.5°C is also the point at which scientists worry that natural systems could spiral out of control in feedback loops and tipping points – the Arctic tundra releasing methane, the loss of krill in the southern oceans, the desertification of the Amazon rainforest, which already looks like it is starting to emit CO_2 instead of absorbing it. These feedback loops threaten to knock our one and only planet off its stable sweet spot for thousands of years and maybe even permanently. *Our one and only planet.*

Here's a short, sobering reminder of how late in the game it is. If we merely let the fossil-fuel machines that are already on this planet live out their natural lives,* they'll produce enough emissions to take us to 1.8 degrees or so.[1] This is why there are advocates for retiring the heaviest and dirtiest emitters first – the coal plants. But even if we retire them early (which we didn't commit to at Glasgow COP26), we are still on track for more than 1.5°C of warming.

---

*   Committed emissions are emissions that will occur because the machine exists and is living out its natural life. A car bought in 2020, for example, will still be emitting in 2040, and a coal plant built in 2020 sadly may still be emitting in 2070.

This is a climate emergency, and it will only become increasingly urgent to radically reduce emissions as time goes on. Any fossil-fired machine, your cars and kitchen stoves included, bought after today (it doesn't matter when you read this, the statement is still true) is incompatible with a 1.5°C world.

We need to get the production of clean options up to scale as soon as humanly possible and try our best to ensure that every new machine brought into existence is a zero-carbon machine, whether it is your car, water heater, stove, gas heater or the local powerplant. This means pretty much all of the machines will need to be electric, and our electricity will need to be generated by renewables or nuclear.

It is urgent, and it is only going to get more urgent.

We've just witnessed the 26th attempt at international climate ambition, COP26, and it wasn't enough. We need faster action. I haven't met the political party anywhere in the world that is moving at the pace we need to. We need to look at every opportunity for speeding up the transition to clean energy, whether that be politically or personally. You'll learn in this book that the no-bullshit way to speed up this switch is to electrify everything.

Why me?

I grew up in Bardwell Park, beside a golf course and a nature reserve, on a stream that might be called a tributary of Wolli Creek. Then it was Sydney's southwest; now, in a city more than double the size, it is grouped with nearby suburbs under the moniker of 'inner west'. My mother was an artist, half of our house was her studio, and she painted and did etchings of Australia's landscapes, waterways, flora and fauna. She was half Mary Poppins and half David Attenborough. My father was a professor of textile engineering, a practical man who also built our house and maintained all the machines in it. He was half Chevy Chase, half Henry Ford. Mine was a great childhood; we had more than enough of everything, especially adventures, as my mother would put pins on the map of the national parks and wetlands we should visit, and my father prepared the vehicles and camping equipment needed to get there. There wasn't any religion in the house other than Attenborough documentaries, so you might call my parents' relationship a match made in the heavenly places of Australia.

I'm writing this book now as a citizen recently returned to Australia after spending two decades in the US. I am a newcomer to the contemporary Australian climate debate, having only observed it from afar for the past 20 years. In the US I have been heavily involved in energy data, and in the energy research and development sector, including building successful technology companies (and some not so successful ones) in wind, solar, natural gas, hydrogen, energy storage and heating ventilation and cooling (HVAC). I am broadly familiar with all energy technologies, have a background in both engineering and physics, and have modelled household, state, national and international energy systems in great detail, including for the US Department of Energy.

I wrote this because I love Australia. I love her wild lands and exquisite yet fragile flora and fauna, and I believe emphatically that Australia has a leading role to play as the world addresses climate change. I believe that this will benefit all of Australia's citizens, as well as Australia's environment. I am writing this of my own volition; I have no ties or vested interests in the Australian energy economy, only an interest in our land and our people.

I earned two degrees in engineering in Australia. The first was a Bachelor of Science in metallurgical engineering from the University of New South Wales, the second a Master's in Civil Engineering from the University of Sydney. My first industrial job was on the 'Rod and Bar' line of the steel rolling mill in Newcastle. My second job was at a Comalco recycling smelter in Western Sydney. I was a Co-op scholar – my degree was paid for by Australian industry, and I am no stranger to coal, steel, aluminium and the extractive realities of Australia's export industries. I worked on products in post-consumer waste recycling before moving to the US to get a PhD at the Massachusetts Institute of Technology (MIT).

The Australian public service is politically cautious in its climate recommendations, understandably so given the decapitations that happen to people who speak up on the issue. In the absence of career public servants, it has become the domain of think tanks, politicians, consultants and corporate Australia, who for the large part represent their interests and give anything but an unbiased view. The Australian Renewable Energy Agency (ARENA) models lots of things very well but has to stop short of 'picking winners', lest it be seen as political or as showing favouritism. By not picking winners, it continues to support ideas that keep fossil fuels going, instead of ending their use completely.

For this reason, the conversation about climate change is typified by fearmongering about jobs and costs and rarely looks ten years out at the big trends and even bigger opportunities. Brave individuals such as Ross Garnaut and Tim Flannery have added some optimism and urgency to the case, with books and concepts such as the 'Energy Superpower' entering the public lexicon. This is the right direction to travel, and now we need even more detail to bridge the imagination gap of the average politician, and to paint a picture for the average punter of how it is going to be better than OK, maybe even great, for them.

Policy in Australia feels like it is set by opinion polls and election cycles – and it now feels we are permanently in election cycles. While our energy system is fundamentally a matter of long-term planning and infrastructure, governments are allowed to make vague commitments and intone that the market will fix it, and that technology, not taxes, is the way forward. This is to shirk responsibility. The government's financing decisions critically influence infrastructure that will last 25–50 years. Government policy on extractive industries will determine the fate of the planet and our climate. We the people, or we the punters, deserve better, forward-looking policies, but we will only get them by demanding them. This book will help you to know what you need to write to your local member about, and how you should vote on energy and climate issues if you care about your kids, your country, your environment, your finances.

There was much noise around the 26th United Nations 'Conference of the Parties' or COP, known to most as 'Glasgow', Australia announced a less than ambitious commitment to net zero by 2050 and hinted at a content-less 'plan' and 'roadmap' to get there that was shrouded in much secrecy. By the time you get to the end of this book you should be qualified to assess the ambition and reality of any politician's or political party's 'plans' to slow global heating. I present my opinion strongly here about the technologies and the approaches that Australia should take to lead the world in decar-bonising a first-world economy. I am convinced we have both the most to win and the most to lose. I also believe that through leadership and example, Australia could not only lead the world but actually increase the ambition of the other nations to clean up their energy acts. Yes, Australia should be lead-ing the US, trouncing the UK and Europe, partnering with China, helping Southeast Asia and setting a shining example for everyone else.

I started this book only a few days before Glasgow was set to begin. I'm finishing just after COP26 ended. I wrote a longer book on mobilising climate solutions targeted at the US president and policymakers, titled *Electrify* and published by MIT Press, which was received well in the US prior to COP26. I will borrow many of those ideas but my focus here is on the Australian context and the peculiarities of Australian solutions. I was typing as politicians announced commitments that are insufficient to save our Great Barrier Reef. I was fighting frustration as world leaders used deceitful assurances and creative accounting to pretend we are on track. I was disappointed by the outcome of COP26, just as I was disappointed in the outcome of COP3, 'Kyoto', which was the first UN climate conference that motivated me to civil disobedience. I was part of a large group of cyclists that blockaded the Sydney Harbour Bridge on the eve of Kyoto to protest Australia's failure to commit to that agreement. We have failed 26 times to produce the necessary ambition through the UN COP process. Insanity is doing the same thing over and over and expecting different results.

People need a vision for climate success that they can gravitate towards, not a fear they recoil from. It needs to be a vision that engages citizens, and with sufficient detail to calm their anxieties. Australia is poised to be the country that can most easily make this transition and show the world a positive success story. That is the gift that we could give to the world: leadership and vision, and an example of success that other nations and their economies can follow.

The former Australian government made a lot of noise about using 'technology, not taxes' and letting the 'free market' solve the problem. But the 'free market' doesn't know what the limit should be on global heating. And on energy and infrastructure issues, the market is anything but 'free'. For one thing, the 'free market' has already built-in subsidies for fossil fuels – whether tax breaks, direct subsidies to corporations, artificially low resource leases or low-interest loans. The 'free market' will not keep us to a 1.5°C target because it is not transforming the energy industry at the pace required. This is why wartime analogies are apt – markets can be influenced by governments during wartime to produce the goods needed to win the war. We need to move faster than the market, and that means that governments must have an opinion and must work towards targets. At this point, 'technology, not taxes' and 'let the market do its work' are clearly code for further delay and inaction. This is not a communist screed! Markets are useful and we should use the best ideas from all sides of the ideological aisles, but most of all we should be guided by what science tells us we must do to avoid planetary calamity – that is, rapidly reduce our emissions. Hazy targets 30 years out are a delay tactic. We should have five-year plans and commitments and hold ourselves to them.

It will take two decades to transform our energy economy. That makes the maths simple. We must move aggressively and we must move now. The market is moving by itself; we aren't standing entirely flat-footed, but we need to get to a sprint while our politicians still drunkenly stumble.

In this book I use publicly available energy and economic data along with analysis tools developed in partnership with the US Department of Energy to fill in the Australian imagination gap. We can succeed much faster than you have been told. It cannot happen without aligning (1) public policy with (2) private industry and (3) the Australian household, which very likely necessitates (4) the financial sector.

Australians are known for their solidarity in the face of adversity. Our RSLs, our surf lifesaving clubs and our volunteer bushfire brigades are testament to our capacity to solve problems together as a community and as a country. Now is the time to unify our country's approach to climate change with the same old-fashioned dedication and hard work. A human generation is often approximated as 20 years. We have one generation to do the job.

What's ahead in this book

In **Chapter 1: The luckiest country**, I'm going to introduce you to the fundamentals of why Australia has the luckiest, easiest and almost certainly most economically beneficial pathway to decarbonise and get to net zero emissions. Electrification is a vaccine for climate change and just as with Covid, Australia should go hard and go early.

In **Chapter 2: Urgency and emissions**, I'm going to be brutally honest about our emissions situation, locally and globally, and why action needs to be much faster than you have been led to believe. Science implies we cannot beat 1.5°C of warming unless we move faster than the free market. In a bar fight between science and the free market, the free market might win, but science will still have been right. It is delusional to think we can get this done on time without designing and influencing the market with incentives, mandates, rebates and subsidies for the solutions that science shows are right.

In **Chapter 3: Energy**, I'm going to lay out the Australian economy in terms of its energy flows, which will help frame the conversation about how we get from here (fossil-fuel dependence) to there (fossil freedom!).

In **Chapter 4: Australia's energy options**, I will introduce you to the things that will and won't get us through this critical energy transition.

In **Chapter 5: Electrify (almost) everything!**, I'm going to build my core argument that you need to forget the hype about hydrogen, lose the idea that everything will run on biodiesel, ditch delusions about carbon sequestration and negative emissions, and instead focus on electrification as the solution. Electrify our homes, electrify our vehicles, electrify our businesses, electrify our industry, electrify agriculture, electrify Australia.

In **Chapter 6: Cheap and getting cheaper**, I'll show you the inexorable trends that are the evidence that we will all benefit enormously from the falling costs of our electric climate solutions.

In **Chapter 7: Electrifying our castles**, I'm going to show you why electrification is going to save you a lot of money, and that we are about to undergo the largest wealth transfer in history from traditional providers of energy to traditional consumers of energy.

In **Chapter 8: Crushed rocks – the export economy**, we'll see why Australian industry and exports can make a huge contribution to the global energy transition while generating far more value and income than our existing fossil-fuel exports.

In **Chapter 9: Why politicians and regulations matter**, we will look at things we've done right on policy, and things we need to do to enable this transition to happen – and to happen at the lowest cost and with the lowest friction.

In **Chapter 10: Financing fossil freedom**, we will look at the enormous importance of finance, both the cost of finance, but critically also the access to finance. These questions strike at the structural decisions that we can make that determine who wins economically with electrification.

In **Chapter 11: So long, and don't kill all the fish**, we'll reflect that it is possible to address climate change while still screwing up some other aspects of our magical and beautiful continent and planet, and suggest that we should go forward boldly, but mindfully.

Finally in **Chapter 12: An abundant Australia,** we will peek at the future that is already happening and remind ourselves that Australia's clean energy future can be abundant and improve all of our lives.

If you are nerdy enough to want to look at scales and energy unit conversions, head to the **Appendix: Scales of energy use** to get a sense of different energy units.

Chapter 1
The luckiest country

- With low population density and extraordinary renewable resources, Australia has the easiest path of any developed country to zero emissions.

- If we are confident and act aggressively, Australia can save money in the suburbs, create jobs in the regions, improve public health, improve environmental and water quality, grow our export economy and address historical inequalities, all while leading the world in addressing climate change.

Donald Horne penned his book *The Lucky Country* in 1964 to describe Australia. His phrase is still commonly used today to capture the good life and relative stability and comfort that many Australians enjoy. But the book was actually a critique; the final chapter began:

> Australia is a lucky country run mainly by second-rate people who share its luck. It lives on other people's ideas, and, although its ordinary people are adaptable, most of its leaders (in all fields) so lack curiosity about the events that surround them that they are often taken by surprise.

Written at the same time as Horne's critical tome was 'Restoring the Quality of our Environment', the 1965 report of the environmental pollution panel of the US president's Science Advisory Committee. It also began critically:

> Ours is a nation of affluence. But the technology that has permitted our affluence spews out vast quantities of wastes and spent products that pollute our air, poison our waters, and even impair our ability to feed ourselves.

Within the report was another report of a subpanel of the committee on 'Atmospheric Carbon Dioxide', which accurately presented the climate challenge we now face. More than half a century has passed since then, and no country has made particularly good progress in dealing with it.

In spite of all that, I remain optimistic, and this book makes the case that we are not just a lucky country, but the luckiest. If we allow our first-rate people to lead, if we believe in our own ideas and ignore false gods sold to us from afar, if our ordinary people continue to be adaptable, it is Australia that can lead the world on climate change. And in taking that leadership role, we will reap the largest rewards not just for ordinary citizens and their energy bills, but by growing and diversifying our export economy with higher value-added metals and items produced using our abundant renewable energy. This will be an export economy that restores our environment and soils, rather than degrading them.

We have everything to win. We also have everything to lose.

What we have to lose is not the nonsense about jobs and industries and lifestyles and weekends and trucks and barbecues – that is the cynical banter of culture war political candidates who want you to believe we'll lose jobs in solving climate change. We won't, we'll create more jobs, and everywhere, including in the communities that currently operate our fossil-fuel industries. What we actually have to lose is our spectacular and fragile ecosystems on this spectacular and fragile continent. The ecosystems that give us life and nourishment. So I'm going to place our spectacular and fragile ecosystems in the what-we-have-to-win column, because we either keep them, and we win, or we lose them.

Win, win, win!

If we go hard and go early on cutting emissions – and if by so doing we encourage other countries to increase their ambition and follow us – we have everything to win. We'll be winning so much, we'll win, win, win, win, win.

The first thing we will win is lower energy prices for all Australians. Driving our vehicles will be cheaper than it has ever been. Heating our homes and our showers will be cheaper than ever. Our electricity will be cheaper, too. The average household will probably save $5000 a year or more on energy and car expenses. There are two reasons for all this winning. One is the rooftop solar miracle. Through good policy and training programs, Australian rooftop solar is the cheapest electricity delivered to residential consumers in the world. This cheap solar energy will power more efficient electric vehicles (EVs), heat pumps and induction cooking, which is where the savings come from. It's a simple recipe.

In winning that first thing, savings in our suburbs, we'll win a second thing: lots of local jobs. Manufacturing, installing and maintaining all the machines that will achieve win number one will require a huge number of tradies redeveloping our national suburban and commercial infrastructure, namely our homes (or castles) and our small businesses. These are the jobs installing rooftop solar, heat pumps, batteries, new appliances in our kitchens, and vehicle chargers; and the jobs insulating and upgrading our homes and businesses. You can't offshore installation jobs because you can't do maintenance on an EV or a heat pump from India, Europe or China.

Win number three will be creating enormously profitable export industries around the things that Australia already does well. Around a third of the cost of making steel or aluminium is the energy required to make it. If you have the cheapest energy in the world you can make the metals that the world needs at the lowest price. Australia has low population density and best-in-the-world solar and wind. We can and will be dominant in cleanly making the things the world needs to get through this energy transition – steel for wind turbines and electric vehicles, aluminium for transmission lines and lightweight structures, copper for all the wiring, lithium and cobalt for the batteries, silicon for the solar cells, uranium for nuclear power. The list is in fact much longer, but you get the point. We win on exports and the jobs they will create in the regions.

Win number four is a win for our health. Unless you have a coal plant in your community (I'm sorry about that), it is the burning of fuel in vehicles, lawnmowers and other engines that has the worst impact on your local air quality (along with the bushfires that they exacerbate by causing climate change). Burning methane and closely related hydrocarbons in our homes (gas-flame stovetops) is a leading cause of respiratory illness. It is just a successful and deliberate PR campaign that has you calling this toxic flammable gas 'natural'. Many of the things you do to improve your health also improve climate outcomes – eating less meat, walking more, cycling more. Everything we do to mitigate climate change will improve our health and quality of life. Our air, water, suburbs, workplaces, schools, churches and libraries will all be cleaner. Even our oceans will be cleaner.

Win number five will be preserving our beautiful places. We might yet save some coral reefs, we'll save many of our forests, we'll regenerate our soils, we'll improve the health of our waterways. The beautiful places we go for recreation should get more beautiful, not less. In learning to care for the whole Earth's climate, we will also have the opportunity to redress our conflicts with the natural environment, to learn from First Nations peoples around the world, and especially the original caretakers of Australia, that we can live well and not in conflict with our land. This is in no way to suggest we need to abandon technological and other progress, but rather that we need to turn our extractive relationship with nature into a partnership with nature.

Win number six is in fixing inequalities. It should be simple enough to understand the statement that you cannot half-solve climate change. You can't have half the people subscribing to, and affording, the solutions, while leaving the less wealthy half behind. In 2021 many people likely feel that they can't afford the technological wonders that are at least a good part of the answer to climate change: electric cars, rooftop solar systems, a household battery, heat pumps. If we choose to actually address climate change we are de facto choosing to help everyone afford those solutions. We can empower households like never before, and in doing so we will need to make sure that households at all income levels can afford to be part of this transition and benefit from this new abundant Australia.

We need to paint more pictures in people's minds of their decarbonised lives being bigger, brighter, better, healthier, wealthier, friendlier and more engaged. Here we go.

Calling bullshit

This book wants you to understand that we can get to a zero-emissions Australia with an 'electrify everything' approach, and that is how we will be best served in thinking about rapid and deep emissions cuts, and climate change.

But a lot of people will be bringing along other ideas, maybe ideas you heard your crazy aunt espouse at the Christmas table. Rather than let you read the 'electrify everything' argument in long form while you still believe some other strategy might work, I'm going to liberate you from some of the bad or impractical ideas now.

Carbon sequestration can't do it

When you burn a hydrocarbon, it becomes three times bigger as a molecule, because you add two oxygen atoms to each carbon atom. It also becomes about 5000 times larger because it becomes a gas (CO_2), where in most cases it started as a liquid or a solid. The fossil-fuel industry is invested in the idea that you can keep burning their product, provided you capture that CO_2 and bury it. To capture the CO_2 requires expensive filtration equipment that, even if you exclude the cost of capital, requires a great deal of energy to operate.

So the idea amounts to using more energy to capture the CO_2, then yet more energy to compress the CO_2, then yet more energy again to transport that CO_2 to somewhere the geological formations allow you to hide it for hundreds or thousands of years while you hope it turns into a rock. Today we pull roughly 10 billion tonnes of fossil fuel out of the ground each year. It becomes 30 billion tonnes of CO_2. The fossil-fuel industry wants you to believe they can keep emitting a whole bunch more because we'll be able to sequester it, but the world is already counting on a physically unrealistic amount of sequestering to compensate for 'overshoot' – the fact that we'll go past our 1.5°C target and have to use negative emissions. In a nutshell, carbon sequestration can only make fossil fuels more expensive, which will make them even less competitive with cheaper renewables. We can't do enough carbon sequestration to maintain even a fraction of the existing fossil-fuel industry. It is unlikely the world will be able to do enough carbon sequestration just to offset our overshoot, let alone to allow for continued offsetting of fossil fuels.

Electrification will be cheaper anyway, and will reduce the amount of energy we need by half while improving our quality of life.

Geoengineering can't do it

Most geoengineering plans are in the world of solar radiation management. This is the idea that we block energy coming from the sun from warming up the planet. Once you go down this path, you have to maintain it in perpetuity. It's a problem even worse than the legacy of nuclear waste disposal. If we do this geoengineering while still pumping CO_2 into the sky, then for some reason we stop the geoengineering because it gets too expensive or too hard, or politically problematic, then all of a sudden we'll experience an extremely rapid spike in global heating because a huge pile of CO_2 will turn the temperature up very quickly. The ocean continues to acidify even if you are managing the solar radiation. We might need geoengineering for reasons unforeseen, so we shouldn't stop researching it, but it is not yet a viable technology and we don't have time to wait. We have the technology to electrify everything; we just need the ambition.

Hydrogen is a dumb way of doing it

Hydrogen is a battery, not an energy source. To make hydrogen with truly zero emissions it has to start with green electricity anyway, only it squanders more than half of that energy as waste in the energy conversion. In short, hydrogen is electrification, but a roundabout and very inefficient way of doing it. If we do hydrogen for the majority of the energy economy, it doubles or triples the amount of clean energy we have to produce. It is an expensive side-show to the main event: electrification. Yes, a small amount of hydrogen, but let's not get carried away. Beware! There are obvious reasons the gas industry likes to lobby for this idea. Because Australia has so much misplaced hope and hype around hydrogen, I'll address this in detail in Chapter 4.

A focus on eliminating waste won't do it

Plastic bags are bad when they wind up in waterways. Too many things are disposable. We throw away a ridiculous number of soft drink cans, take-out containers and coffee cups. These things are all true, and worthy of our attention, but recycling plastic and reusing shopping bags has a negligible effect on the warming climate. Nitrous oxide emissions from plastics production do have a climate effect, and we should look for alternatives, but this is not the main event. Yes, reduce, reuse, recycle, but most of all cleanly electrify and eliminate emissions. If we all use Keep-cups, stainless-steel water bottles and canvas tote bags, we won't even come close to fixing climate change. We will hardly move the needle. In contrast, if we electrify everything, we'll be most of the way there.

Veganism can't get us there

Reducing meat and dairy consumption is one of the easiest personal choices people can make to reduce their climate impact, specifically cow products (beef and dairy) and sheep products. But alone it will not solve our climate problem, as agriculture makes up less than 20% of global greenhouse gas emissions. The largest proportion of emissions, by far, is our energy use, which is why we need to electrify and decarbonise how we use energy.

There are a number of problems with meat. One is the amount of land required to grow the feed. Another is that ruminants (cows, sheep) belch methane, which is far worse as a greenhouse gas than CO_2. On an infrastructure scale, better land management and new low-carbon farming alternatives will lower the impact of eating meat occasionally. We don't have to give meat-eating up altogether, but it does need to become more conscious. Meat and dairy alternatives are growing wildly in popularity and quality, which will help to further reduce the impact of meat consumption on the climate.

Overpopulation is not the problem

Many people seem still to believe that the world population is just growing in an upward direction, won't stop, and that this is a major problem for humanity. A recent survey of 10,000 young people aged 16 to 25 found that around four in every ten are hesitant to have children because of climate change. This misconception and population panic is definitely not limited to young people, and we should clear it up. Population experts predict the world population will level off at around 10–11 billion people.[2] To provide some context, we currently have around 7.9 billion people. From an energy perspective, enough sunlight hits Australia to supply the world's current energy needs many times over, and that's just one form of renewable energy. Where will all these people fit? The website 'Wait But Why' entertainingly explored some quick calculations for this, showing that 7.3 billion humans standing shoulder to shoulder could fit into a square about 27 kilometres on each side. That's 729 square kilometres. The greater Sydney region is 12,368 square kilometres, and Tasmania is 68,401. So yes, the entire world population from every continent could stand shoulder to shoulder and fit into just one Sydney suburb. Obviously this is no way to live, but it illustrates just how much space there really is in the world. Of course there's more to this story than just space and energy – but world population growth is not something to get incapacitated by. We know that the recipe for lowering population growth and improving quality of life involves providing better access to education for young women. We can lower the peak population number by advocating for and investing in women.

Living like people used to won't do it

The romantic notion of our global ancestors living 'in balance with nature' is not supported by evidence. It turns out, as detailed beautifully by Hans Rosling in his book *Factfulness*, that our ancestors 'died in balance with nature', with such high death rates that the impact on nature was less per person. To quote the book, 'it was utterly brutal and tragic'.[3]

Life expectancy, globally, has more than doubled in the last 200 years alone (from 31 years in 1800 to 72 years in 2017). In those same 200 years, the proportion of children who die before their fifth birthday has been reduced by a factor of ten (from 44% to 4%), and the share of adults worldwide with basic reading and writing skills has gone from 10% to 86%. In the last roughly 90 years, plane crash deaths per miles travelled have been reduced by a factor of over 2000, and in the last roughly 20 years, extreme poverty worldwide has been reduced to around half what it was. Humanity has made a lot of positive progress in a short period of time. This is not to say that anything today is perfect; we have a lot of work to do. We have things to learn from the past, but looking at our past with rose-tinted glasses won't help us to build a better tomorrow for humanity.

What about nuclear?

There are still plenty of cranky old (mostly white) men, who just by chance seem to mostly have physics and or nuclear engineering degrees, who continue to posit that we cannot solve climate change without nuclear energy. The world can solve this crisis without nuclear, and with renewables alone, but the task might be made easier with some nuclear. This is a 'yes, and' solution. Yes: some nuclear is good. And: we also need lots of solar and wind and hydro and heat pumps and electric cars and batteries. Australia doesn't need nuclear because of our mild climate and incredible resources, but we do have a lot of uranium to help other countries who need to take this path – countries where darkness or cold or high population density make it a challenge to rely only on renewables.

Renewable energy is more than capable of doing it

Australia can make renewable energy in far higher quantities than our own economy needs. We can also store renewable energy for tomorrow. We can make so much that we have enough in the dark cold times and an abundance in the warm, sunny times, and a surplus to sell to other nations.

Growing trees won't do it

Growing trees is a fantastic way to address climate change. Even if we grew 1 trillion of them, which is ambitious to say the least, it would only have a small impact on climate change. By all means pay someone else to plant some for you, or even better plant a few thousand yourself. I did as a teenager; it's not hard and it's very rewarding. But to really move the needle you also need to electrify everything in your life, including the car that drives you to the land you plant those trillion trees on.

Carbon taxes (alone) won't do it

A carbon tax isn't a solution. A carbon tax is a market fix meant to motivate all the other solutions to compete. It's designed to slowly increase the price of carbon dioxide, and to slowly make fossil fuels uncompetitive. The idea is that a high enough carbon tax would make all the fossil fuels more expensive than at least some of the other solutions, which in a perfectly rational market would prompt people to use those solutions.

Carbon taxes might have been sufficient if we'd started with them in the 1990s but for the taxes to achieve the 100% adoption rates we need now, they would have to ramp up very quickly, like tomorrow. They would be difficult to implement, as well as regressive, hitting lower-income people hardest. It is probably just as effective to eliminate fossil-fuel subsidies, which in many markets would tip the scales in favour of alternatives anyway. And by the time we have the political will to implement a carbon tax, renewables with batteries will be cheaper than fossil fuels.

A carbon tax is useful in decarbonising the hard-to-reach end points of the material and industrial economy, but alone it won't be rapid enough to transition home heating from furnaces to heat pumps, and vehicles from internal combustion engines (ICEs) to electric vehicles at the rate required.

What about the impact of making all the new things?

It is worth considering what we might do wrong in trying to address climate change. Remember that 'it was a good idea at the time' applies to coal, oil, natural gas and nuclear. Some people sweat about whether we have enough of all of the materials to make this clean energy transition. Some worry about the waste. With certainty there will be unforeseen consequences, but for a moment we should focus on what we know. Today, the average Australian uses around 6 tonnes of fossil fuels per year, 23 tonnes if we include our exports. Even just those 6 tonnes are a lot and translate to around 17 tonnes of CO_2 emissions per person per year.

But what if we did it differently? The same lifestyle, fully electrified, could be had for 4 kilowatts (kW) of constant power, which is 96 kilowatt-hours (kWh) per day, per person. If we were to commit to producing half of that with solar and half with wind, and – because it's not always windy or sunny – store half of it in batteries, we would need far less stuff. How much less? Assuming the wind turbines will last 30 years, the solar modules 20 and the batteries ten, all reasonable numbers for technology we have today, we'd need around 50 kilograms per year per person of wind turbines, solar modules and batteries – around 150 kilograms in total. Wind turbines are made from a lot of steel and copper and some glass and plastic. Solar modules are mostly glass and aluminium with a little silicon. Batteries are made of lithium and other elements that are quite recyclable. We should be able to recover as much as 80% or more of these materials. If we aggressively recycle these sustainable options, we would need just 15–25 kilograms of basic stuff per person per year, a hugely smaller burden on the planet than the 6000–7000 kilograms we produce in the carbon economy today, measured in raw carbon or combusted CO_2, practically none of which is recycled.

There is reason for hope here. Fifteen to 25 kilograms is about 300 times less material than 6000 kilograms. A thousand times less than our individual CO_2 contributions. This analysis is plainly back of the envelope, but in it you should find hope, optimism. We can be far better stewards of the planet.

Can we make enough batteries?

No two ways about it, we will need a lot of batteries. This is not impossible, however, given current levels of manufacturing capacity. To replace Australia's 20 million (and the world's 1 billion) personal gasoline-powered vehicles with electric vehicles in the next 20 years, we will need over a trillion batteries, or around 60 billion 18650 batteries every year (18650s are

18 mm in diameter and 65 mm long – slightly larger than your flashlight's AA-size battery). That is fewer than the 90 billion bullets manufactured by the world today. If you need only one statistic to summarise what is wrong with humanity, it is that we only make about 19 billion LEGO bricks every year, yet we make 90 billion bullets – enough to shoot everyone on Earth eleven times a year! Imagine the world where we made 90 billion LEGOs and cut our bullet consumption back to just a few billion. We need lots of batteries, but it is possible. We need batteries, not bullets.

Beware of the Doomers

There's a frightening amount of climate fear and anxiety, especially in our youth. There's also no shortage of doom and gloom from older generations. Doomers come in all ages. Pessimism leading to inaction will not help, and more importantly, we have the opportunity for a far brighter future than Doomers think. In many different ways, this is the best time in human history to be alive. We have work to do, especially on climate, but with the right action we can continue to make the best time to be born in human history today, every day, as we move into the future. Let's focus on building that brighter future. When people try to overwhelm me with negativity, I typically call them out: 'OK, Doomer. Do you have a better idea?'

A lucky time in a lucky country?

There are strong reasons to believe that we might be about to launch the most transformative movement in history – a movement that redefines our extractive relationship with the planet, lightens our footprint and improves our lives. It might just be the luckiest time, in the luckiest country.

Chapter 2
Urgency and emissions

- Only a wartime-style emergency response will avoid climate calamity.

- Machines that already exist will burn far more than enough carbon to take us past 1.5°C of warming.

- We must move more aggressively than free-market economics can achieve.

The spirit of this book is to look forward and find solutions, not to look backwards and lay blame. Before looking at what fossil fuels have caused, let's take a moment to thank fossil fuels for getting us here. We didn't know the harm they'd do when we started using them, and certainly they have helped to lower infant mortality, double life expectancy, cut extreme poverty, decrease manual labour, increase quality of life, and a whole host of other things that bring us progress and prosperity. In 1896 and even earlier we knew carbon emissions might be a problem,[4] but broad public awareness of climate change and carbon dioxide didn't really arise until NASA scientist James Hansen testified about it to the US Senate in 1988.

We know we need energy, but by now it is clear that we can no longer have CO_2 emissions as the by-product. Unemotionally presented, the recent history of human-caused emissions is shown in Figure 2.1. The industrial revolution and the discovery that coal could drive steam engines really kicked it off, and the US, Britain and Europe rode those fossil emissions to prosperity.

2.1: The rise in CO$_2$ emissions from fossil fuels and cement production from 1800 to 2019, and the countries they are attributed to.

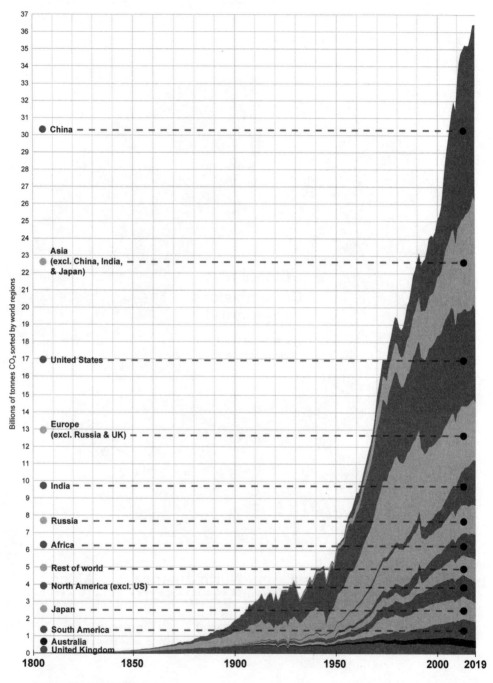

Source: Global Carbon Project

We can see how these emissions have accumulated in our atmosphere, by country of origin, in Figure 2.2. The US and Europe are responsible for much

2.2: Cumulative total CO_2 emissions (1750–2019) by country.

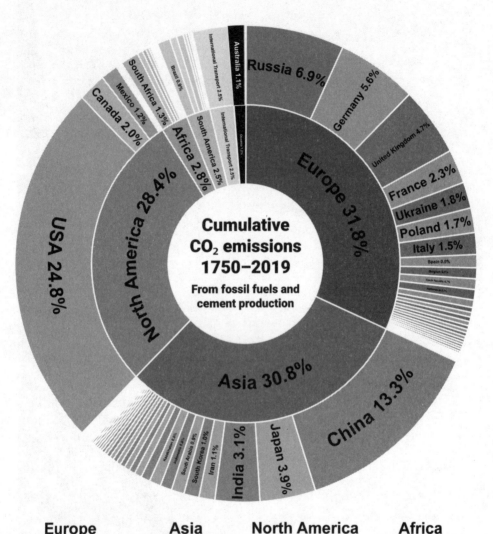

Europe
31.8% of global emissions.
13.1 billion tonnes CO_2

Asia
30.8% of global emissions.
12.7 billion tonnes CO_2

North America
28.4% of global emissions.
11.8 billion tonnes CO_2

Africa
2.8% of global emissions.
1.2 billion tonnes CO_2

South America
2.5% of global emissions.
1.0 billion tonnes CO_2

Intl. transport
2.5% of global emissions.
1.0 billion tonnes CO_2

Oceania
1.2% of global emissions.
0.5 billion tonnes CO_2

Source: Global Carbon Project

THE BIG SWITCH

of the emissions so far. Australia is responsible for 1.1% of what is already in the air, three to four times our pro-rata (or per person) share given our population (around 0.33%). But in fact Australia is responsible for much more than 1.1%, something more like 3–4% if you include the emissions from our coal and gas that are exported overseas and burned elsewhere.

But that also is too simplified a picture. Emissions by country in 2019 are shown in Figure 2.3.

China is now the dominant emitter, although many of China's emissions are exported as products that people in the US, Europe and Australia buy. The future's big emitters are likely to be India, Brazil, Indonesia and Africa (as well as the US and China). If these countries take the same fossil-dependent path to development as the US, Britain and China, there isn't much chance of any climate outcome you'd want to live in. If they develop sustainably with renewable energy, nuclear power and end-use electrification, we have a hope.

Together these pictures tell us that while North America and Europe are largely responsible for emissions to date, it is what China, India, Brazil and Africa do next that is going to be the big determinant of our collective climate outcome. Unfortunately Figure 2.3 and Figure 2.2 are used to apportion blame, or to justify this emission or that one, whereas both fail to express that we all live in the same atmosphere, and the outcome is dependent on all of us. As Carl Sagan put it, 'Carbon dioxide molecules are exceptionally stupid. They don't know anything about national boundaries . . . the national boundaries have no bearing on these global environmental issues, no one nation can solve this problem by itself.'

Blame isn't very useful. Action is. We are in a race for more ambition and a path to a better future. If Australia demonstrated a pathway to zero emissions that was abundant and aspirational, we might expect a lot of the world to follow. A developing nation that can't see a better way forward might justify walking in the fossil-dependent footsteps of the past – but if a shining example of prosperous sustainable living existed, we might have

2.3: 2019 global emissions by country. Australia is responsible for 1.1% as they are counted, or 3.4% (dotted lines) if we count the burning of exported Australian fossil fuels.

China 23.5%

United States 11.8%

India 6.6%

European Union (27) 6.4%

Russia 4.8%

Indonesia 4.5%

Brazil 2.8%

Japan 2.6%

Iran 1.8%

Thailand 0.8% | Turkey 0.8%

Pakistan 0.8% | Italy 0.7%

Vietnam 0.6% | Spain 0.6%

Egypt 0.6% | Kazakhstan 0.6% | Ukraine 0.6% | United Arab Emirates 0.5%

Poland 0.7% | France 0.7%

Colombia 0.5% | Algeria 0.4% | Cameroon 0.4%

Venezuela 0.7% | Tanzania 0.6%

Democratic Republic of Congo 0.5% | Myanmar 0.4% | Bangladesh 0.4% | Iraq 0.4%

Germany 1.6%

Canada 1.6%

South Korea 1.3%

Australia 1.1%

South Africa 1.0%

Paraguay 0.4% | Uzbekistan 0.4%

Angola 0.4% | Peru 0.3%

Austria 0.1% | Azerbaijan 0.1%

Mozambique 0.2% | Oman 0.2%

New Zealand 0.1% | North Korea 0.1% | Papua New Guinea 0.1% | Serbia 0.1% | Zimbabwe 0.1%

Botswana 0.1% | Cambodia 0.1% | Finland 0.1% | Hungary 0.1% | Bahrain 0.1% | Congo 0.1%

Ethiopia 0.4% | Netherlands 0.4% | Malaysia 0.3% | Bolivia 0.3% | Sudan 0.3%

Mongolia 0.1% | Portugal 0.1% | Ireland 0.1% | Nepal 0.1% | Ghana 0.1% | Honduras 0.1%

Zambia 1.0%

Nigeria 1.0%

Philippines 0.3% | Turkmenistan 0.3% | Czechia 0.2%

Afghanistan 0.2% | Libya 0.2% | Morocco 0.2%

Kenya 0.1% | Laos 0.1%

Mexico 1.4%

Saudi Arabia 1.3%

Argentina 1.0%

United Kingdom 0.9%

Ecuador 0.2%

Mali 0.1% | Bulgaria 0.1% | Guinea 0.1% | Niger 0.1%

Greece 0.2%

Sweden 0.1% | Somalia 0.1% | Sri Lanka 0.1%

Kuwait 0.2% | Belgium 0.2% | Chad 0.2%

Qatar 0.2%

Israel 0.2%

Belarus 0.2% | Uganda 0.2%

Syria 0.1%

Australia 1.1%

Australian fossil fuel exports (extra emissions if counted)

Source: Global Carbon Project

more hope that they could bypass pollution in favour of something obviously more beneficial to their citizens.

Australia needs to stop blaming the actions of other countries for our own inaction and for our fossil profiteering. Someone has to lead.

It should be us.

Emissions trajectories

Figure 2.4 dramatically shows us the urgency of heavy emissions reductions in the decade from 2020 to 2030 if we are to achieve the 1.5°C of warming widely seen as the upper limit we shouldn't cross if we don't want to throw the climate into (even more) disarray. To have any hope of a 1.5°C degree world without banking on lots and lots of negative emissions, we need to reduce emissions by almost 75% by 2030. Every year we delay makes this harder. If we don't act decisively this decade (which really means this year), our chance of limiting warming to 1.5°C will be gone.

2.4: Emissions reduction pathways needed to keep the world below 1.5°C of warming without counting 'negative' emissions.

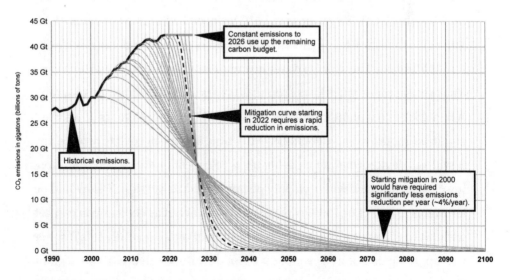

Source: Robbie Andrew, Global Carbon Project, IPCC and Raupach et al. 2014

To give themselves and their citizens the illusion that we are doing better than we are, many governments are banking on future 'negative emissions', and lots of them. Figure 2.4 shows clearly the problem with relying on sequestration at unprecedented scale. Each year that we delay, the required rate of reduction goes from steep to precipitous. If we don't make any cuts by 2026, 1.5°C cannot be achieved without an unrealistic quantity of negative emissions. It boggles my mind that we are betting our one and only planet on a technology not yet proven, at a scale larger than any industry that has ever existed.

Let me try to be even more concrete about this, as this is a practical book striving to be frank. Figure 2.5 places the 1.5°C and 2°C emissions trajectories on the same graph (a). The second part (b) of the graph contemplates a fossil fuel–powered machine that lasts 20 years, and the effect of the adoption rate of clean alternatives on the stock of machines over time. At 100% adoption rate, every machine is replaced at the end of its life, and so after 20 years all the machines have been replaced by clean technology. If the adoption rate is only 50%, after 20 years half the machines have been replaced by new technology, half by fossil-fuel machines that will continue to emit CO_2. To stay within our climate window we need adoption rates of over 50% starting NOW. To get close to 1.5°C we need 100% adoption rates ASAP. Part (c) of the figure highlights that, in practice, different machines have different lifespans. The average water heater or stovetop lasts around ten years. The average car or truck lasts around 20 years (before it is turned to scrap). A new gas or coal plant installed today will last around 40 years. If we were about as ambitious as we could be, every time anyone needed to replace any fossil-fuel device, whether a gas heater, a petrol car or even a power station, it would be replaced with a cleanly powered electric version. This would be known as a perfect or 100% replacement rate, about as good as you can get without tossing out machines before they wear out. Hopefully this'll sober you up. If we do about as well as we can do, and we electrify those giant consumer industries immediately, and we switch all electricity supply to

2.5: (a) 1.5°C and 2°C pathways with no negative emissions (b) 10, 25, 50 and 100% adoption rates for machines with 20-year lifetimes (c) perfect replacement of machines with lifetimes of 10, 20 and 40 years in the context of 1.5°C and 2°C emissions trajectories.

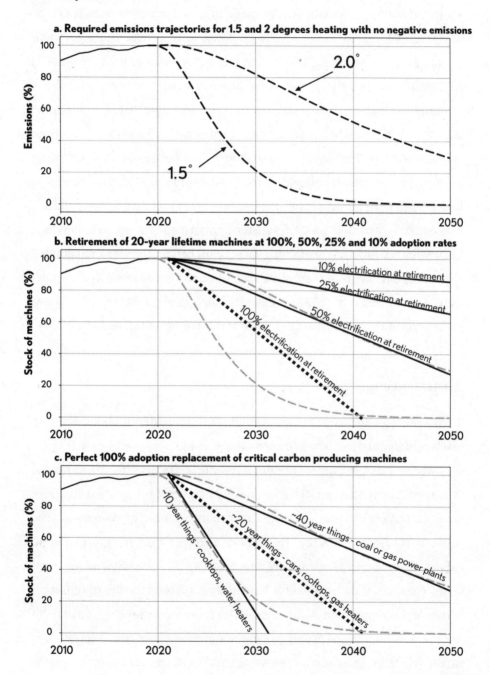

renewables as soon as possible, and then everyone else in the world follows suit, we'll still pass 1.5°C. That's why we need an emergency, wartime-style effort. That's why we need to retire the heaviest-emitting coal plants before their end of life. That's why we also need to invest in all the negative emissions technologies we can – as well as, not instead of, electrifying everything.

Figure 2.5 hopefully says it all. This is in fact an emergency. It does in fact require an emergency response, not haphazard waiting for the market or god or the flying spaghetti monster to provide. Try reconciling that reality with 'net zero by 2050', which to date is the best the Australian government has been able to commit to. The linguistic contortions of our politicians are spectacular, and one can only imagine they hope that you either (a) find it all too confusing and will just trust them to keep on doing 'something', (b) are too stupid, busy, exhausted or distracted to care, (c) are too uninformed to notice that they don't know what the fuck they are talking about or (d) believe the same bullshit that their lobbyists have sold them about clean coal and gas-led recoveries. I don't think we are going to get change unless the Australian people, en masse, know better, expect better and hold governments and representatives to task.

Australian emissions

The reason to reform our energy system is to lower our emissions. But first we should understand where our emissions come from, and where they go. All governments are required to submit inventories to the Intergovernmental Panel on Climate Change (IPCC) according to set categories. The categories are not always easy to decipher, but the accounting mostly makes sense. The problem for Australia is that this reporting doesn't cover exports. For an export-heavy country such as ours, that creates a distorted view of our emissions. This distortion has not only hindered progress on the decarbonisation of our local economy, but also concealed Australia's enormous opportunity to become a zero-emission exporter of energy-intensive products. Australia can in fact help the world decarbonise while building the industries of the future.

2.6: Australia's emissions breakdown as reported for 2019 (summarised). (LULUCF includes emissions associated with land use, land-use changes and forestry.)

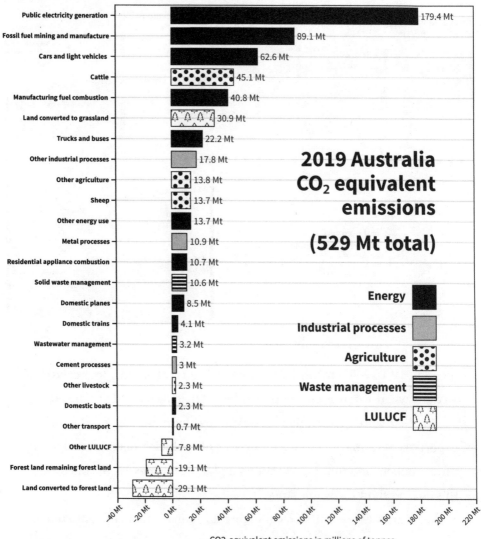

2019 Australia
CO$_2$ equivalent
emissions

(529 Mt total)

Public electricity generation	179.4 Mt
Fossil fuel mining and manufacture	89.1 Mt
Cars and light vehicles	62.6 Mt
Cattle	45.1 Mt
Manufacturing fuel combustion	40.8 Mt
Land converted to grassland	30.9 Mt
Trucks and buses	22.2 Mt
Other industrial processes	17.8 Mt
Other agriculture	13.8 Mt
Sheep	13.7 Mt
Other energy use	13.7 Mt
Metal processes	10.9 Mt
Residential appliance combustion	10.7 Mt
Solid waste management	10.6 Mt
Domestic planes	8.5 Mt
Domestic trains	4.1 Mt
Wastewater management	3.2 Mt
Cement processes	3 Mt
Other livestock	2.3 Mt
Domestic boats	2.3 Mt
Other transport	0.7 Mt
Other LULUCF	-7.8 Mt
Forest land remaining forest land	-19.1 Mt
Land converted to forest land	-29.1 Mt

Energy
Industrial processes
Agriculture
Waste management
LULUCF

CO2-equivalent emissions in millions of tonnes

Source: Australian government Paris Agreement inventory 2019

Figure 2.6 shows that Australia's largest emissions are from our electricity generation, which is dominated by coal. Fossil-fuel mining and manufacturing (not even burning it, just finding it and preparing it for burning, either in Australia or overseas) is our second-largest source of emissions. Our cars and

2.7: Total emissions that emanate from Australia if we also include fossil fuels that we export that are burned overseas.

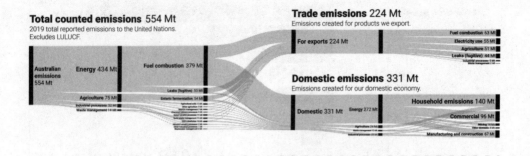

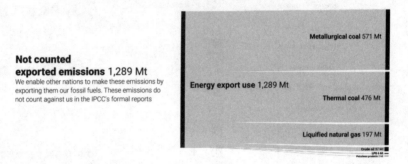

utes and 4WDs roll into third place, followed not far behind by our cows. Not all the cows are eaten by us – more than half of that meat and dairy is exported.

In other words, Australian emissions are dominated by producing our exports, generating our electricity, driving our vehicles, making our agricultural products, and running our homes and small businesses. But once again, this doesn't show the whole story.

We redraw these emissions in Figure 2.7 to make them easier to understand, using a Sankey flow chart* of the emissions that emanate from Australia. Only about one-third are formally counted as Australia's under the IPCC accounting criteria – a total of 554 million tonnes, known as megatonnes (MT), per year. Critically, we see here that around 40% of the emissions that

---

*   Named for Captain Phineas Rhiall Sankey, a Scottish sea captain who invented this type of flow diagram to describe energy transformations in his steamship.

count against us (224 MT) are emissions incurred in creating the things that we export. All these exports are mined and farmed using diesel fuel, coal, natural gas and electricity. The remaining 331 MT – our domestic emissions – arise from the things that we do in our domestic economy: driving, heating and cooling our homes, running our businesses, and creating the food that we consume here in Australia. Just on the numbers, as Figure 2.7 shows, we are actually responsible for far more emissions. The burning of our coal and other fossil-fuel exports creates emissions that dwarf those of our domestic economy. Around 1290 MT of emissions are embodied in our exports – more than double the emissions that formally count as ours, but these are nonetheless emissions that we enable people in other countries to make. Looked at this way, roughly 22% of our emissions are for us, while 78% are for other people.

Australians could reduce our domestic emissions to zero and still contribute heavily to climate change by continuing to export countless tonnes of fossil fuel. Cynical politicians argue that Australia should not have to cut emissions as much as other countries because these emissions aren't ours and are actually in service of other economies. That is to admit we are arms dealers, selling the weapons (fossil fuels) that will be used against our children and future generations. What we're missing is the opportunity that such an export-driven economy presents. In a world constrained by climate change, where we are one of only a handful of countries with ample, if not extraordinary, renewable-energy potential, the world will be hungry for decarbonised supply chains. Australia could still be first to offer zero-emission steel, aluminium, copper, other metals and agriculture.

Emissions at home, emissions abroad

Australia exports more than 75% of our coal and around 80% of our gas. We export around 70% of our total agriculture production, including 75% of our beef, 73% of our sheep meat and around 71% of our wheat.[5] We export close to 900 million tonnes of iron ore, yet we only make around 5 million tonnes of steel domestically. We mine more than 100 million tonnes of bauxite, but only convert a tiny fraction into 1.6 million tonnes of aluminium.

We are led to believe that exports are good for the country, but whose country? Perhaps Australia shouldn't take responsibility for the emissions of these emissions-intensive export industries because they aren't actually owned by Australians. More than 80% of our mining production is foreign-owned. BHP and Rio Tinto, which the public think are good Australian companies, are more than three quarters foreign-owned. Arguments abound about who is responsible for what, and these disagreements are a major cause of delay on climate action. Pushing responsibility this way or that determines who wins and who loses, making the politics extraordinarily difficult, as can be seen in the failure of your average COP to put us on an honest track.

Separating our emissions and our exports allows us to refocus a conversation that is usually mired in the fear of losing export dollars. We can think about what can be done to decarbonise our domestic economy, which is a political issue felt by households and small businesses; this will be covered in Chapter 7. Separately, we can think about how we can continue to have a thriving export economy while reducing the emissions of that export economy to zero. By creating zero-emissions exports we can generate revenue for our country and help the rest of the world get to zero too. We will look past the currently ill-advised and culturally polarising conversation about exports and the Australian economy in Chapter 8.

Chapter 3
Energy

- The energy we use in our daily lives is the main source of our emissions. Most of this energy starts its journey as coal, an innocent little black rock.

- We can drastically reduce our emissions by replacing fossil-fuel-burning machines with electric machines. This includes supply-side machines, like power plants, as well as demand-side machines, like gas water heaters.

These days, people's awareness of energy as an issue is driven by our concerns over climate change. Our emissions of CO_2 and other greenhouse gases, including methane and nitrous oxide, are the problem. Globally, around 75% of our emissions come from energy. The other large contributors are agriculture (11%), land use and forestry (6%), industrial processes (6%) and waste (3%). Many of these other emissions are indirectly caused by the underlying use of fossil fuels to serve those sectors. Ending fossil fuels as the basis of the world's energy supply is by far the most significant thing we can do to get to zero emissions.

The 20th century was so effective in developing fossil-fuel infrastructure that most people are completely oblivious to the amount of energy consumed in their daily lives. When I moved to Wollongong with my wife in 2020, we passed over a rail bridge as a mile-long coal train travelled beneath. She asked, 'What is that?' I thought everyone knew what a coal train looked like, something so critical to their modern lived experience. I had to explain that sadly, the majority of our electricity comes from coal. My wife was later bewildered, when walking in the bush behind our house, to learn that the black rocks poking out of the ground were also coal. The country is made of the stuff, I had to remind her; she had only never seen it in the US because she'd never looked.

Australians, just over 25 million of us, consume about 125 million tonnes of coal a year. That's 5 tonnes per year, or 14 kilograms a day, each. There is nothing else in your life that you consume at this rate, and in second place is the petrol and diesel we put in our cars. It isn't obvious when we fill up our cars because we don't actually watch the precious liquid go into the tank, but each Australian household uses around 2800 litres per year of these liquid fuels, which is a little more than 1000 litres per person per year, or 2.7 litres per day. Along with your eight cups of water a day, you also 'consume' about eleven cups of petrol, or the average person does. In 2018 we used 1200 kilograms per person per year, or 3.3 kilograms per day, of natural gas, an energy source that is literally invisible to us.

If you put on a backpack full of all the fuels you needed to get through the day as an average Australian, it would weigh about 20 kilograms. Every day we burn those 20 kilograms of fossil fuels, which turn into 60 kilograms of CO_2. In this day and age it is pretty hard to hide dead bodies, yet each of us basically hides a dead body-weight of carbon every day. It is amazing that we've hidden this from ourselves for so long . . . something sinister so easily achieved with a few swipes of a credit card and some auto-deposits.

Australia keeps relatively good energy statistics, structured similarly to those of the US Energy Information Administration (EIA). These are published annually as the Australian Energy Update, a summary of which is presented as a Sankey diagram in Figure 3.1.[6] The supply side (where we get our energy from) is on the left, and the demand side (what we do with that energy) is on the right. The energy inputs to the Australian economy are clearly marked, and we can see the dominance of black coal and natural

3.1: Australian energy use, 2018–19, from the Australian Energy Update 2020.

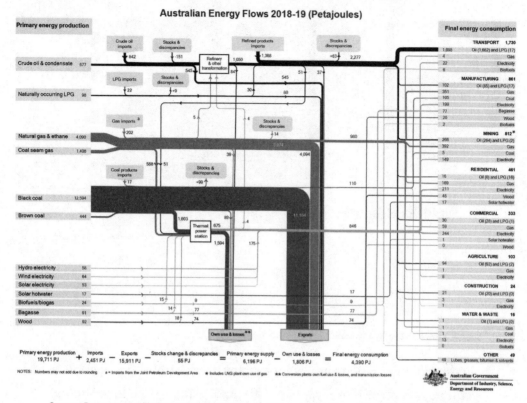

Australian Energy Flows 2018-19 (Petajoules)

Source: Department of Industry, Science, Energy and Resources © Commonwealth of Australia 2021

gas. Our heavy reliance on oil imports for transport is obvious. Imports and exports enter and leave through the top and bottom of the chart. The energy use is further broken down into sub-categories: transport, manufacturing, mining, residential, commercial and services, agriculture, construction, 'other' (there is always an 'other' – the bucket where anything too hard to categorise is placed) and exports. The limitations of these categories are quickly apparent. Transportation includes activities as diverse as international air travel, international shipping, freight trucking and household driving. Some of the energy used in transport actually provides the energy we use in the manufacturing, commercial and residential sectors.

In Figure 3.2, I have created a more detailed Sankey diagram for Australia's energy flows. Immediately one can see the absolute dominance of

3.2: A more detailed Sankey diagram of Australian energy uses.

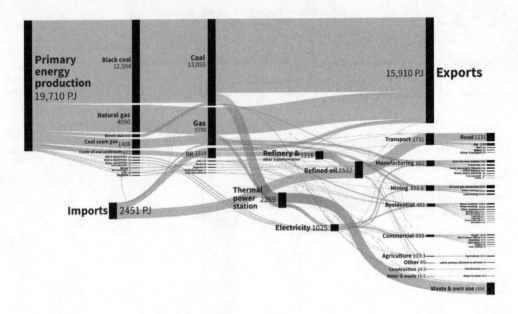

Source: Australian Energy Statistics 2020 and Rewiring Australia

our exports. Energy flows in a large economy like Australia's are enormous, so they are described in petajoules (PJ). A PJ is 10^{15}, or a million-billion joules. In a 21,500 PJ economy, 15,000 PJ are exported. Most people probably don't think a lot about energy units, and those who do probably think in kilowatt hours (kWh). A petajoule is 2.8 billion kWh. It is so many kWh that people also use the term terawatt hours (TWh). A TWh is 1 billion kWh. Our huge energy exports are why the group Climate Analytics highlights that Australia alone could contribute 10% of the remaining emissions in the Paris agreement's implied carbon budget.[7]

One way to change this picture is to export a different type of energy – perhaps electricity, perhaps hydrogen, perhaps ammonia, perhaps something not yet invented. The other way, which we'll find out more about in Chapter 8, is to export our energy in a different type of crushed rock. Today we sell millions of tonnes of coal. Tomorrow we could be selling millions of tonnes of our metals and ores, mined, processed and transported with our renewable energy. We'll also find out that these technologies aren't exactly working yet. There are fairly clear pathways to making zero-emission steel, aluminium, ammonia and other high-energy materials, but these things are unlikely to be ready to scale up until the late 2020s or early 2030s, so let's turn our attention to what Australia can do *today* while we also support our scientists, engineers and entrepreneurs to do the innovation for what we need *tomorrow*.

Domestic economy energy flows

As can be seen in Figure 3.3, our domestic energy economy needs a supply of around 7000 PJ today. This supply is met with 1800 PJ each of coal and natural gas, and around 2400 PJ of refined oil. Electricity isn't a source of energy; it is a type of energy. Generating electricity consumes the most energy of any activity in the Australian economy – the great majority fed by coal and

3.3: Domestic energy flows when we exclude exports.

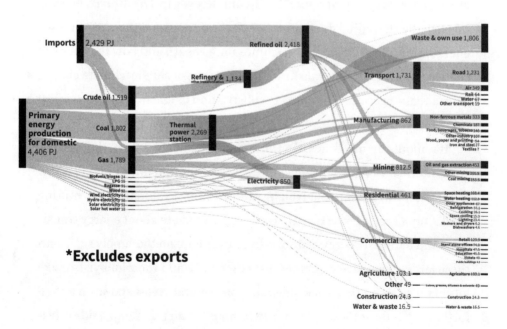

*Excludes exports

Source: Australian Energy Statistics 2020 and Rewiring Australia

natural gas. Of the roughly 2300 PJ of supply used to generate electricity, around 1800 PJ is wasted as heat in thermal power stations.

Moving to the demand side of the picture, where we use or 'demand' energy, more than 1700 PJ are used in transportation, two-thirds of that on roads, and 350 PJ in air travel. A very large portion of this is wasted in the inefficient little internal combustion engines in our cars, which are only about 20% efficient at converting energy into taking us places. Space heating, water heating, cooling and cooking are the biggest energy consumers in our buildings, be they residential or commercial, and much of this is supplied by natural gas. Our manufacturing industry is energy-intensive; similarly, mining uses a huge amount of energy, much of it to mine fossil fuels.

I am an enormous fan of Sankey diagrams, as you can probably tell. I've spent years staring at them, making them and understanding them. You probably have crossed eyes and are skipping over the pages with these spaghetti charts. So in the spirit of practicality, in Figure 3.3 I've drawn cartoons over the top of the Sankey diagram to give you a picture of the machines underneath. Maybe you remember Keanu Reeves in *The Matrix*, peering through the streaming numbers to see the reality behind them. When I see energy data I think about the machines underneath and how they need to change. This is how you transform the challenge of solving climate change from a vague exercise to a concrete substitution of machines that practical people can understand.

It's about the machines. On the supply side of this chart, the left, is a small number of enormous machines. A few hundred power plants, a few thousand mines, some big freight rail machines, some tankers, mining trucks, an oil refinery or two. Historically, the debate about energy transitions and national energy plans has lived over here on the supply side, with these large, expensive machines. Big companies and their lobbyists make lots of noise about these machines and long capital cycles and so on and so forth. We'll look at the options for switching out all this supply-side infrastructure in the next chapter.

3.4: True insight into the energy economy occurs when you see the machines underneath the energy flows and start thinking about how we are going to replace them with something better.

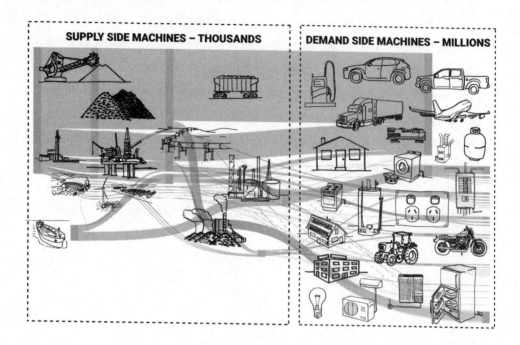

SUPPLY SIDE MACHINES – THOUSANDS

DEMAND SIDE MACHINES – MILLIONS

In the middle are all the means of transporting and transmitting this energy around the country. Oil and gas pipelines, freight rail for coal, transmission and distribution lines for electricity. Supply chains for LPG, tanker trucks and petrol stations for our cars and trucks. A substantial amount of energy is used in keeping this supply chain running – as much as 10% of our fossil fuels are used just to keep the fossil-fuel system going.

On the right-hand side of this chart, the demand side, is where we, the Australian people, spend every day. Our lives are entwined with the very large number of small machines that use all this energy. Our vehicles, our air conditioners, our toasters and stoves and hot tubs and pool pumps and water heaters and Nintendos and laptops. Historically, these machines have not really been a strong part of the climate debate – perhaps because they bring us too close to the reality that we are all complicit in our energy use and emissions. These machines are all ready to be replaced. Many of them are electric already (laptops, fridges, and so on) but crucially, many of the big ones still run on fossil fuel (water heaters, space heaters, cooktops, cars). This demand side is where the action is for the next decade.

It is a truism that supply equals demand in the energy world. Yes, we need to swap gas turbines for wind turbines and coal plants for solar plants (the supply side), but that won't get us all the way. We also need to swap out all the machines on the demand side. Later in this book, we'll learn that we have the option to do this immediately, and that the technology is ready. We'll see why it will save us money and improve our lives, our environment, our air and our water. But first, to achieve this grand vision, we need to look at what our options are for generating all this energy without generating emissions.

Chapter 4
Australia's energy options

- Australia is a vast, sparsely populated country with diverse and abundant clean energy resources, far more than we need.

- Solar will be the dominant form of renewable energy, followed by wind.

- Hydrogen, much hyped, has a role to play – but probably a small one.

No matter how you look at it, harnessing energy is a land-use issue. This is true whether we are digging coal mines, fracking natural gas deposits, mining uranium ore, installing massive wind turbines or giant arrays of solar. The world is roughly a sphere. It's not getting bigger. Most of the planet is populated. To think big-picture about our energy system requires thinking big-picture about land use.

Current Australian land use

Figure 4.1 lays out pretty starkly the division of use of our wide brown land. The overwhelming land use in Australia is grazing. Almost all of this is grazing sheep and cattle, which explains why emissions from our agricultural sector are so high. We have given half of the country over to possibly the

4.1: Land use in Australia.

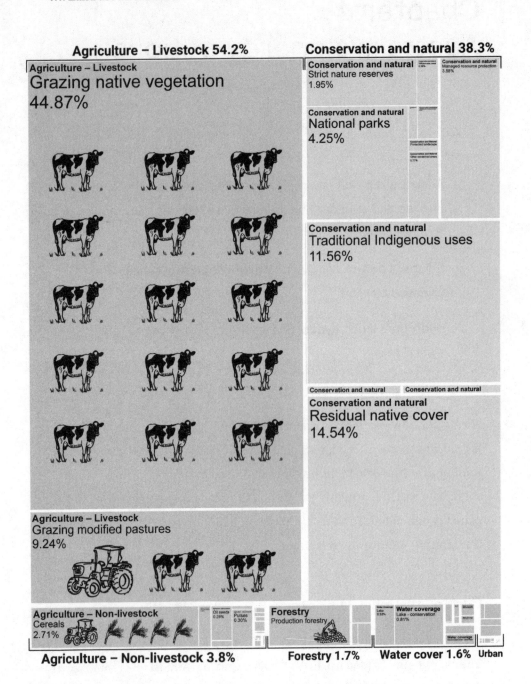

Agriculture – Livestock 54.2%

Agriculture – Livestock
Grazing native vegetation
44.87%

Agriculture – Livestock
Grazing modified pastures
9.24%

Agriculture – Non-livestock
Cereals
2.71%

Oil seeds
0.28%

Pulses
0.30%

Agriculture – Non-livestock 3.8%

Conservation and natural 38.3%

Conservation and natural
Strict nature reserves
1.95%

Conservation and natural
Managed resource protection
3.58%

Conservation and natural
National parks
4.25%

Conservation and natural
Traditional Indigenous uses
11.56%

Conservation and natural

Conservation and natural

Conservation and natural
Residual native cover
14.54%

Forestry
Production forestry

Forestry 1.7%

Water coverage
Lake - conservation
0.81%

Water coverage

Water cover 1.6%

Urban

Source: ABARES 2016

worst land use in terms of emissions production: ruminant grazing. I like eating the odd cow, enjoy dairy, and like any good Australian kid love a roast or spit lamb, so I'm not blaming farmers or omnivores or carnivores here, just stating plainly that this is an incredible amount of land to dedicate to something we know is destructive. It represents an enormous opportunity for Australia to innovate and the CSIRO is making good progress, including with seaweed dietary supplements that lower cows' emissions.[8]

The land we devote to roads is another interesting example to keep in mind. I was once lucky enough to ride the Indian Pacific railway from Sydney to Perth. Ours is a big country. Not many of us experience these huge expanses of land on a daily basis, but anyone who has driven between two Australian capital cities will have a feeling for how many roads there are in Australia – almost 900,000 kilometres, or approximately 1,800,000 'lane kilometres'. If each lane is 3.5 metres wide, that's 6300 square kilometres of road surface, or about 0.08% of the country.[9] Keep this image in mind, as we'll come back to it shortly.

A global perspective

One of Australia's blessings as a potential renewable energy superpower is our low population density combined with our hugeness. With so few people per square kilometre, and so much land, we have the opportunity to produce more energy than we do currently and more than we need domestically.

In Figure 4.2 I have mapped the world's countries according to population density on the left axis, and total landmass on the bottom axis. Australia is one of a small group of very large countries that includes Russia, Canada, China, the US and Brazil – and we have the lowest population density of all of them. India, although large, is in a fairly distant seventh place. We have a lot of land that can be used to generate renewable electricity, relative to our population. Australia and Brazil have warmer climates with more consistent solar radiation year-round, which gives them a further advantage compared to Russia and Canada, where the winters are brutal and dark.

4.2: Population density (people per square kilometre) vs land area (millions of square kilometres). The right axis shows the approximate percentage of land use needed to power each country's energy needs with renewables.

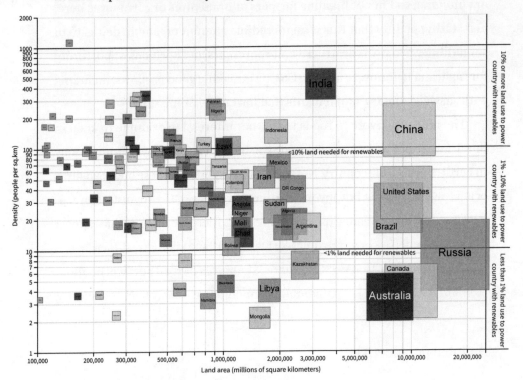

Using existing fossil fuels, citizens of high-income countries (Australia, the US, Norway) consume around 10,000 watts per person. As we will find out, powering the same lifestyle with electricity backed by renewables will halve that to about 4000 to 5000 W per person. This assumes living like we do today, only electrified. In Europe, experiments by the 2000 Watt Society have shown that 2000 W per person can easily power a comfortable life with more thermally efficient homes and more widespread use of public transport, cycling and walking.

We know that a good solar module can produce around 40–50 W per square metre (on average, all day, every day, through the whole year). To give us that comfortable 4000 to 5000 W electric lifestyle, some very basic maths tells us that each person needs the equivalent of 100 square metres of solar, a square ten metres on each side. If your population density is ten people per square kilometre, simple maths again tells us that around 1% of the land area of the country will need to be covered in renewables. If the population density is 100 people per square kilometre, you'll need to cover 10% of the land. Most countries fall somewhere in between. To its great good fortune, Australia has 3.35 people per square kilometre. We need to cover less than one third of 1% of our land with renewable energy generation.

You might see where I am going. Australia is huge, we have a low population density, and we have an advantageous place relatively close to the equator with great solar. As a general principle, I'm going to suggest that countries with fewer than ten people per square kilometre have the opportunity to be net renewable energy exporters, particularly if they also have a mild climate (harder for Russia or Canada). For countries with a higher population density, it is very likely that they will either need to use nuclear power or will end up importing energy from a country with a lower population density. Competing land uses such as agriculture will make it difficult for more population-dense countries to dedicate significant land to renewables.

In short, Figure 4.2 tells you why Australia has the easiest path of any country in the world when it comes to being a renewable superpower. Lots of land, lots of sunshine, not a lot of people. It also emphasises why nuclear may be necessary for some very high-density countries.

Australian energy resources

It is hardly a coincidence that if you have a lot of land, you have a lot of resources. Australia's spectacular abundance of energy resources – fossil, nuclear and renewable – is well documented. We'll look at renewables and nuclear in more detail here to see how we can get to zero. Quite a few studies of how Australia can run on 100% renewables reach the same conclusion: it is not only possible but can be done cheaply and quickly.[10] Let's look at the options we have for powering our great nation.

Solar

Solar energy is most typically farmed either with photovoltaic (PV) modules – in other words, solar panels – or with what are called 'solar-thermal' power plants. Averaged throughout the year – a single orbit of the sun – the sun hits the outer atmosphere of Earth with around 1386 watts on every square metre. Only about 1000 watts per square metre get through the atmosphere, measured directly above Earth at about noon. Averaged throughout the day, and throughout the year, compensating for cloudy days and atmospheric losses and even dust, the average incident solar radiation

4.3: A map of Australia's world-class solar energy resources (horizontal surface solar irradiation).

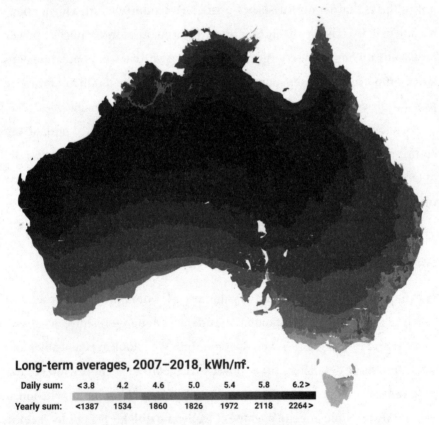

Long-term averages, 2007–2018, kWh/m².

Daily sum:	<3.8	4.2	4.6	5.0	5.4	5.8	6.2>
Yearly sum:	<1387	1534	1860	1826	1972	2118	2264>

Adapted from World Bank, Global Solar Atlas 2.0, 2020. Solar resource data : Solargis

on the ground is around 300 watts per square metre (W/m^2). A typical, high-quality, modern PV module is just above 20% efficient at extracting energy, meaning it captures one fifth of this energy, or about 60 W/m^2.

It is difficult to cover 100% of a space with solar because you need walking space between modules for cleaning and so forth, so let's assume 50% ground coverage. Typically the inverter that turns the captured energy into the 240 volt electricity that we like is about 95% efficient. This means we get, very approximately, 30 watts per square metre of energy on land covered with a solar farm. If we use this very rough number, we find we get 2.6 MJ per square metre each day (about two thirds of a kilowatt, which is 3,600,000 joules or 3.6 MJ). This adds up to 700 MJ per year per square metre which is a number we can use to have some fun. The phenomenal Australian solar energy resource can be seen in Figure 4.3.

Just for laughs, let's imagine that Australia used all its grazing land for solar farming, instead of cow and sheep farming. After counting up all the zeros (there are a lot of them), Australia would produce nearly 3 billion PJ 450 times the total domestic demand of 6500 PJ, 200 times Australia's energy exports, and five times the entire global primary energy need (or about ten times the global need if everything was electrified). That is to say, we have solar in abundance. If we reduced our meat output by 20%, we could supply the whole world with energy on that rangeland. That idea might ruffle some feathers at the sausage sizzle, but it sure would beat 4°C of climate change.

We could look at this in another way. An area equivalent to all our roads would produce 4400 PJ. That's enough to power our whole electrified economy.

As everyone has observed, we have seasons. Melbourne has (at least) four of them, the rest of us maybe just two, summer and winter, or in the case of Darwin, bloody hot and even hotter with rain. These phenomena can be seen in Figure 4.4. The summer wet season dampens summertime solar production in Darwin, but for the rest of the year it easily outshines the other capitals with its high northern latitude. Darwin has a nearly constant

4.4: Daily solar energy in Australian capital cities, kWh per square metre per day.

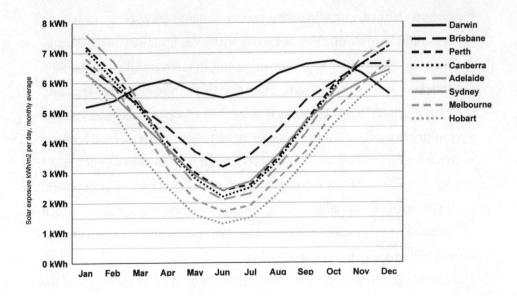

Source: Bureau of Meteorology

year-round supply of 6000 W/m², which works out to around 3 kWh per square metre of rooftop per day. Tasmania obviously has the weakest solar irradiance, given its proximity to Antarctica. While it does briefly peak above Darwin in December and January, in wintertime the output reduces fourfold from the summer peak. Solar is still viable in Tasmania, but this output highlights the need for seasonal storage, although Tasmania's spectacular hydroelectricity resource puts it in a great place to be the first state in Australia to be 100% renewable. The rest of the country sits somewhere between these extremes, with a fantastic summer resource that reduces by about a half to two-thirds in the depths of winter.

Solar thermal

Some people still champion the two technologies known as solar thermal. One is the iconic Australian 'Solahart' domestic hot-water heater, which

THE BIG SWITCH

heats water with sunshine heating a metal collector on the roof. The other is industrial solar thermal, which, using mirrors, concentrates solar energy by about 1000 times and focuses it onto a tower or receptacle, where the intense heat is used to heat a molten salt, which in turn can be used to run a steam turbine to generate electricity. Despite much early hope for solar thermal (I was briefly entranced in the early 2000s), it looks like the majority of solar installed will be photovoltaic (PV). Solar thermal is less tolerant to clouds and atmospheric dust and dirt, takes up more land, and – because the generation is ultimately a heat engine – suffers the fundamental efficiency losses of conversion. It can in theory provide 24/7 power because heat can be stored in the salts, but after many decades it still does not compete economically. Even the rooftop solar water heaters are likely to prove less popular than covering the same roof area in PV. Running a heat-pump water heater powered by PV achieves similar or even higher efficiency and allows more of your energy system to be electric, which makes balancing all your other electric loads throughout the day easier. Said another way, it is hard to get full utility out of the area dedicated to heating the water, because hot water can't also run your car – but if you are using electrons for both, and you have batteries, you'll get more out of every centimetre of rooftop.

Rooftop solar

A study of rooftop solar capacity estimates we can produce 245 TWh with our roofs, which is more electricity than the entire Australian grid delivers in an average year.[11] This number will go up with improving PV technology, but already represents around 900 PJ.

Australian rooftop solar is the envy of the world. A small band of advocates on the south coast developed installation techniques in the early days that they translated into certification and training programs, which greased the skids for low-cost installation. The underlying modules cost

0.25 cents per watt, and wind up being roughly $1 per watt installed, which is a kind of magical number where the electricity costs around 5–7 cents per kWh – cheaper than the cost of transmission and distribution on a regular network.[12] For the regular grid, if you include transmission costs, distribution and metering, the cost averages 12–13 cents per kWh, not including the cost of generating the electricity. This is momentous, because it means rooftop-generated electricity will now always be your cheapest option, even if nuclear fusion or some fantasy technology were free! And if batteries get to ten cents per kWh per cycle of storage, which is predicted as soon as 2024, then the economics will forever be in favour of installing as much local rooftop solar and as much battery storage as you can fit. Want to know why traditional energy suppliers are trying to slow this trend down? Because it kills the business model that was so profitable for so long.

Yet some of the studies are still underestimating rooftop potential; here are three reasons why it's even better than we think: (1) solar will be so cheap that we'll start using north-facing vertical surfaces as well as our rooftops; (2) solar continues to increase in efficiency – its efficiency is now often 20%, up from 14% or so only a decade ago, and will likely eventually top out at around 30%, which means more energy generation per square metre of your precious rooftop; and (3) shade-resistant PV modules are in development and will soon enter the market, making the losses caused by trees and the shade of other buildings at certain times of day less of a problem.

Wind

The world has 80 TW of available wind power 100 metres above the ground, where typical turbines reach.[13] There is 380 TW if we look at the wind potential in the jet streams, ten kilometres high. This wind at altitude is partly the reason I started a company called Makani Power in 2006 (ultimately purchased by Google and run as a Google X project until 2019) to cheaply capture all that energy. In short, wind is abundant globally, and could safely supply half of the world's energy in a 100% renewable energy world. Australia has particularly good wind resources, especially on the western and southern coastlines. We have names for this resource – the 'Fremantle doctor' and 'the roaring 40s', as we call the strong winds circulating the globe at 40 degrees of longitude and wreaking havoc on every second Sydney to Hobart yacht race.

Table 4.1 lists what one study sees as a viable energy mix for Australia in 2050.[14] The main takeaway is that experts agree that the biggest portion of the load will be met by wind and solar. Wind energy has matured enormously since the 1980s and is now considered the cheapest electricity source of all in many parts of the world, cheaper than electricity that can be generated by coal or gas. Turbines have consistently grown bigger and bigger, with a typical one now an enormous 80–150 metres in diameter, generating 3–5 MW. Most industry observers expect their output to grow to as much as 8–15 MW, with research groups around the world focusing on how to build 'floating' platforms that can anchor to the seabed in water 50 metres deep or more. Offshore placement of wind turbines gets around some of the land limitations mentioned earlier.

Table 4.1: Modelled approximate percentage of energy supplied by each resource for 2050 Australia. (Source: Jacobson et al., *Joule* 1:1 (September 2017).

ENERGY SOURCE	% OF TOTAL
Onshore wind	20.07%
Offshore wind	16.73%
Wave	5.79%
Geothermal	0.28%
Hydroelectric	3.13%
Tidal	0.10%
Residential PV	13.73%
Commercial & government PV	11.17%
Utility PV	18.11%
Concentrated solar power (CSP)	10.88%

When I started my wind energy company, Makani Power, in 2006, the cost of wind power globally was above 20 cents per kWh. In preparing our pitch to investors we modelled confidently that we would get the technology to three to four cents per kWh and that it would take about a decade. By the time a pilot was put in the water in Norway in 2018, Makani was still on track for that cost target, but the global industry had reduced its cost faster than we could catch up. Wind now costs as little as about four cents per kWh. There are a lot of lessons in there, and one of them is that it is hard to bet against the incumbents in the energy industry, and that the cost reductions of getting to scale can be enormous. Wind energy's learning rate, or the rate at which the cost decreases for every doubling in size of the industry, is about 12%. Globally, wind power is going to double in capacity three or four more times, so we can expect the costs to fall dramatically further – probably by nearly half.

Hydroelectricity

Internationally, hydroelectricity is the largest supplier of renewable energy at the present time, at 4300 TWh, or 15,500 PJ, in 2020.[15] Australia is the driest continent on Earth, but nevertheless has a reasonable hydroelectric potential of 60 TWh per year, or 216 PJ. The current installed capacity is around 8 GW, in 108 facilities, and in 2018 Australia generated around 58 PJ, or roughly a quarter of what we might ultimately achieve if we invest heavily in this technology. Hydroelectricity can be environmentally destructive to the local area that is flooded to build the reservoir it requires, but it is otherwise very reliable and quite environmentally benign. The much-touted Snowy 2.0 project will add 2 GW of capacity, and 350,000 MWh of storage. That might sound like a lot of storage, but if all 20 million vehicles in Australia had 75kWh batteries, they would have a storage capacity of 1,500 MWh, over four times as much – which is why electrifying our vehicles and having vehicle-to-grid technology is so impactful.

Wave and tidal

Wave and tidal energy are broadly classified together under ocean energy, which also encompasses salinity gradients and ocean thermal gradients. None of these technologies has been deployed at scale, and the economics remain unproven. Australia has spectacular tidal resources in the north of the country, and incredible wave resources along the west coast, south coast and particularly around Tasmania. People who have experienced a strong southerly swell or surfed even 12-foot waves will know that the ocean can be brutal. On top of all that, the environment is enormously corrosive. It is a compelling challenge, but I think a betting person would place only a small, long-odds bet on wave or tidal energy becoming anything more than a niche component of Australian or global energy systems.

Tidal machines come in two main categories: (1) underwater turbines that look a bit like wind turbines, which sit on the seabed in the tidal plain; and (2) something like a dam in an inlet that is filled and drained a couple of times a day. Wave-power machines come in all kinds of varieties: some swing masses, or turn cranks, or push air or water in and out of turbines. If you are fond of building better mousetraps, there are plenty of opportunities here.

Finally, there just isn't as much wave and tidal energy as you might think. If you captured all the tidal energy in the oceans (effectively stopping the tides going up or down), you would only supply one fifth of the world's energy needs and wreak havoc doing so. Similarly, if you captured all the waves hitting every coastline in the world (tragically rendering the surf breaks of the world smaller, or gone), you would also only generate about one fifth of the world's power. There is a lot more energy in deep ocean waves, but the challenge of extracting that energy and getting it back to land economically is difficult, to say the least.

Nuclear

About 10% of the world's electricity is currently produced with nuclear power. In the US around 100 reactors produce about 20% of electricity; in France it is 70%, and in Sweden it is close to 40%. All these plants run on traditional designs developed in the 1960s and 1970s. These are mostly light-water reactors that use uranium as the fuel. The uranium is brought to the threshold of a critical reaction; the heat it creates is used to make steam, which spins a turbine that turns a generator that produces electricity. Australia has the world's largest 'Reasonably Assured Resources' of uranium – equivalent to around one third of the known finite resource. There is very likely more uranium in Australia yet to be discovered. Measured in known uranium reserves, the world has about 50 years of nuclear energy at the current rate of utilisation – significantly less if we ramped up more nuclear power.

Australia is also believed to have 10–25% of the world's thorium resources, which is another pathway to nuclear fission. There are some advantages to the thorium cycle, including the fact that there is more thorium than uranium in the world, and that it is only weakly radioactive. China is currently commissioning an experimental thorium reactor.[16]

Nuclear fission, which is the tearing apart of atoms, may supply us with a short-term pathway out of climate change, but it won't power civilisation for 1000 years. For that we need fusion, the smashing together of atoms, which still does not work, despite recent progress, although it should become a viable technology this century. Multiple companies have predicted that they will have demonstration plants within the decade.

Technically, Australia could power itself quite easily, and for a very long time, with nuclear energy, but there has never been public appetite for the perceived risks. Measured in deaths per unit of energy provided, nuclear has an astonishingly good track record, even when we take into account high-profile accidents such as Fukushima and Chernobyl. Compared to the

10 million deaths a year attributable to the air pollution caused by fossil fuels, nuclear energy looks like a tremendous idea.[17] But I think it is unlikely that Australia will go nuclear, for reasons of public opinion. Our recent commitment to American nuclear submarines certainly increases the chances a little bit. But the reality of modern distributed solar – the solar on our rooftops – is that even if nuclear energy were free to generate, it would still cost more than rooftop solar because of the costs of transmission and distribution.

As an engineer, I would be comfortable with Australia developing domestic nuclear-powered electricity generation. Even if only 10% of our electricity came from such a stable and predictable source, it would make it much simpler to meet seasonal peaks and baseload power. Going 100% nuclear is probably not possible because of the amount of cooling water required – that is, if we were to try to use precious freshwater. In the US, 41% of all fresh water is used to cool nuclear, natural gas and coal power plants, and ultimately it is access to cooling water that limits the amount of nuclear power that can be deployed with current technology. There are programs to work on 'dry cooling', but nothing yet commercially viable at scale. The other option is to cool the reactors with seawater, which adds a little expense and scares some members of the public, especially after Fukushima.

The good news for Australia is that it doesn't need nuclear energy to meet its energy demands; it can easily get by on cheaper wind and solar. For highly populous nations, it is harder to be self-sufficient on renewables alone, which is why we can expect a number of our Southeast Asian and Pacific neighbours to invest in going nuclear. Australia has a long profitable business ahead in mining uranium to help the rest of the world get to zero emissions.

Biofuels

Australia could do a lot with biofuels. Already, because of sugar cane, we do.[18] In the US, the excellent *Billion-Ton Report* from the Office of Energy Efficiency and Renewable Energy suggests that the US could produce around

10–20% of its energy from agricultural waste, food waste, human waste and forestry waste.[19] Globally, biofuel, or bioenergy, is important, comprising as much as 10% of global energy supply. Sadly, much of it is burned for heating and cooking in the developing world, which is neither the best use of this precious resource nor particularly good for people's respiratory health. Biofuels comprise around 2% of transportation fuels and about 1.5% of electricity generation globally today. The big sources of bioenergy are concentrated by-products of agriculture. In Australia, the most plentiful of these is bagasse from the sugarcane industry; in the US it is 'black liquor', a by-product of the wood, paper and pulp industry.

Biofuels can compete with food production for land, and certainly the heavily subsidised US corn-ethanol industry looks like a very bad way to create biofuels. In Australia, wood, bagasse and biogas (from sewage and landfill) already contribute more than 200 PJ. With comprehensive collection programs, Australia could easily more than double this, and domestic bio-waste could provide more than 10% of our future energy supply. As biodiesel or jet fuel or similar, this could easily satisfy our aviation industry, our long-haul trucks and mining vehicles and other long-distance transportation. As a winter fuel, firewood is still reasonably economical for meeting heating loads on a few cold days a year, but the respiratory effects in homes are known to be bad. Heavy firewood use in high-density residential areas can lead to low-quality outdoor air quality, too.

The challenge with biofuels is the low energy density – at 10–18 MJ per kilogram of dry mass, it is about half that of coal or a quarter that of oil. This low energy density and its broad geographic distribution mean that it is expensive to collect because you have to transport much larger amounts of it, over longer distances, to get the same amount of energy as other fuels. A profound contribution would be small-scale biorefineries for producing distillate or a similar fuel in small quantities from these distributed resources.

Waste-specific biofuels, be they from food, forestry, farming or human waste, could make a very meaningful impact on the so-called 'hard to decarbonise' sectors. This area is currently under-invested. Biofuels are very likely to out-compete hydrogen in many of these sectors.

Geothermal

Geothermal energy is fairly simple to understand. Heat energy trapped in the centre of Earth from when the planet formed, from a little bit of radioactive decay and – believe it or not – from 'solid Earth tides', or the squishing of Earth by the gravity of the moon in its orbit – all these create heat in the ground that can be tapped. Holes are drilled deep into the surface of Earth and water pumped down. As the water heats up, it becomes steam that can spin a turbine. Sadly, there is not enough geothermal energy in the world to supply growing global energy needs – it represents about 20 TW, the same as the current world energy consumption. That said, in places with the right geological conditions, it can provide stable, reliable and cheap electricity. Iceland benefits enormously from geothermal, which provides 65% of its primary energy (another 20% from hydro makes Iceland almost 100% renewable).

Australia cannot supply all its energy with geothermal, but it could get close. It is unlikely to be cheaper than wind or solar, but it has the advantage of being available all day, every day – meaning that, like nuclear, it could provide baseload power. Some systems for harnessing geothermal require a lot of water because not all the water is recovered. In Australia we would need to promote recycling or closed-loop systems that reclaim all or most of the water.

One challenge with geothermal is the cost of getting the electricity from the plant to where it will be needed. Most of the best geothermal resources are quite remote and require dedicated new transmission lines. A 500-kilometre transmission likely adds 1–2 cents per kWh to the cost of the electricity.[20]

Storage and demand response

The wind doesn't always blow. The sun doesn't always shine. There is winter. There is summer. In a world of renewable energy, we will need energy storage. There are second-to-second, daily, weekly and seasonal storage challenges. The second-to-second problem is the challenge of electricity: the supply, or the generation, must always meet the demand, or the load. This is necessarily true at every instant, every second. If the load starts to increase more than the supply, the voltage can drop. If the load drops and the sun comes out, the voltages can spike. These voltage spikes can lead to brown-outs and blackouts. The daily problem is matching the solar energy that is produced in the middle of the day with energy demands that occur at other times of day. There are weekly problems: a rainy week of low solar production, or a week of no wind due to some unseasonal synoptic system. There are seasonal problems, such as the fact that we like heat in the winter, just when solar produces the least power, and we love air-conditioning in the summer.

For these challenges we need storage or demand response. Storage consists of machines or devices that, like batteries, can stash energy for later. Demand response, rather than storing energy, moves the load –the hot water can warm up earlier in the day when the sun is shining, or the refrigerator compressor can turn off for a while because the door is closed and it can coast through a few more hours without needing to be cooled further.

In the likely case that around half of Australia's energy will come from solar and most of the other half from wind, we are going to need a lot of storage. Fortunately, our cars and homes (including the appliances and heating systems inside them) represent opportunities for lots of storage.

Batteries

Batteries are the hot new thing in storage. They are no longer the AA cells of our childhoods made of zinc, manganese, potassium, graphite, nickel, copper, lead and various acids; today, they are mostly lithium. Lithium is the third element on the periodic table after hydrogen (H) and helium (He), and the lightest of the metallic elements. It is desirable for batteries because it is abundant and lightweight relative to the amount of energy it carries. Lithium got its start in computers, cell phones and toys, but in the past decade has found its big market in electric vehicles, for which most lithium batteries are now destined. Lithium batteries have fallen precipitously in price, from over $1000 per kWh in 2010 to close to $150 per kWh in 2020. Organisations such as Bloomberg New Energy Finance project the cost of batteries will reach around $75 per kWh by 2025.[21] For a complete battery pack – including battery, battery management system (BMS), inverters, safety features and packaging – the cost is widely expected to soon drop to $100 per kWh at the factory gate, and perhaps $150–200 per kWh when embedded in a product. This is not the reality in 2021, where a battery on the side of your house probably costs $1000 to $1500 per kWh.

Why does cost matter? We don't care so much about the cost per kWh as much as the cost per kWh cycle. This is the cost of the battery per cycle of storage, measured in cents per kWh. A Tesla Powerwall in 2021 comes with a ten-year, roughly 3700-cycle warranty. To achieve these long lifetimes, perhaps only 80% of the battery capacity is used. This makes the maths pretty simple – we divide the cost by the useful component of the battery, and divide that by the cycle life, to give us an estimate of the cost per kWh of storage: $1000 \div 0.8 \div 3700 \approx 34$ cents per kWh.

This is why we are on the cusp of a revolution, but not quite there yet. At 34 cents per kWh, these storage costs make electricity very expensive. If you pair it with solar at six cents per kWh and you need to store half of the energy for later, the total cost of providing 24/7 energy will average out to 23 cents per kWh. This is almost but not quite competitive with the grid in most places. If you are in a remote area, it is likely to be economic.

But in 2022 Ford will ship an electric version of the F150 pick-up truck – the F150 'Lightning', the most manufactured 'ute' in the history of the world – for around US$40,000, including something like a 100 kWh battery. Rounding up and converting to Australian dollars, that's around $500 per kWh for the battery, and it comes with a free truck! The point being that once we get to under $500 per kWh – and we should eventually get to $200 per kWh – a solar and battery combination will be a slam dunk against the existing system, with a 24/7 price of electricity of 10–15 cents per kWh and a cost of storage under ten cents per kWh and perhaps as low as five cents.

There are other types of electro-chemical batteries under development, including flow batteries and grid batteries that are literally made of rusting iron dust. There are also new lithium chemistries coming that use less cobalt and other expensive and rare additives. On the horizon, we can squint and see the emergence of 'solid state' batteries, which will be more robust than current lithium cells and have higher energy density.

Energy matters because it determines how heavy your batteries are and how far your electric vehicle can drive. Energy density currently stands at around 250 Wh/kg (which is 1 MJ/kg compared to diesel's 45 MJ/kg). There is some optimism that batteries will achieve 10 MJ/kg, which with the lighter motors of electric drivetrains and efficiency in the 90% range will easily outperform any vehicle with an engine and a petrol tank, which will be limited to around 20% efficiency of conversion.

Heating systems

Hot-water and home heating systems can both be used as inexpensive storage. The water for tomorrow's shower can be heated with today's sunshine very effectively. For hydronic (that means water carries the heat) heating systems, heating reservoirs of water that can be stored for later also creates very cost-effective storage.

A well-sealed home is in and of itself a form of battery. If you heat the home up to a toasty 22°C, it will take a few hours for it to dip below 20°C, when you might wish to heat it up a little again. This is known as thermal inertia and many building practices, including the Passivhaus standard, use these techniques to make very comfortable homes. In effect we can heat and cool our homes in advance and use thermal inertia as a form of storage or demand response. For managing the second-to-second, hour-to-hour and day-to-day electricity fluctuations, this is a highly underappreciated asset that will be useful in balancing the future grid.

Pumped hydro

One very cheap and effective battery is pumping water uphill and letting it generate power on the way back down, just as with traditional hydroelectricity. This requires 'head', which means one reservoir that is higher than another. Australia is an old continent without a huge number of mountains, but there are still plenty of coastal sites where this age-old battery can help us with the week-long balancing problem of supply and demand. Hooking up Tasmania, with its extraordinary hydroelectric resources, to the grid on the continent is an excellent idea. I live underneath the escarpment in Wollongong. It sits about 800 metres above sea level. It is sites like these that offer perfect conditions for pumped hydro, and the proximity to Sydney could be a huge boon to Wollongong, providing stable power and grid services to the people of greater Sydney and NSW.

Hydrogen

Hydrogen should really be included under storage, as ultimately it is a storage medium for energy generated by something else – but given the hyperbole over hydrogen in Australia, it is worth addressing in a little more detail.

If you don't think about it too hard (which is the trend these days, it seems), hydrogen looks like a perfect energy solution. It is the most abundant element in the Milky Way and quite abundant on Earth. Pure hydrogen has an extraordinary energy density of around 120 MJ/kg, which is more than double petrol's 46 MJ/kg. On Earth, however, hydrogen is not abundant in a form that can be used for energy. Pure hydrogen is reactive, so on Earth it is mostly present as a compound, bonded with oxygen (as water), with carbon (as hydrocarbons), or with both carbon and oxygen (as found in petroleum). To be useful in energy systems it first needs to be separated into a gas. As a gas at ambient pressure, pure hydrogen isn't very energy-dense –around 0.09 kilograms or 11 MJ per cubic metre For comparison, petrol stores around 36,800 MJ in one cubic metre. So hydrogen needs to be compressed, which takes energy to achieve. If you've ever felt your bike pump after inflating a tire, you will notice it gets hot; this is the unavoidable wasted energy of compressing a gas. Because compressed hydrogen needs to be stored at high pressure, it requires a tank, and tanks capable of safely containing a highly flammable gas at explosive pressures cost a lot of money. Hydrogen can be transported, but there are challenges. It can be burned to make heat, or used in a fuel cell to create electricity – but there are probably not enough cheap catalysts for the latter. Overall, for reasons I'll explain below, hydrogen in general is not a very efficient way to store or retrieve energy.

Blue, green or grey? What does it mean?

Australia has been particularly prone to the hype around hydrogen. It has been sold to the workforce as requiring the same skillset as our existing natural gas industry, but trust me – as someone who has worked on hydrogen safety – it is in a whole different category. There are incentives for the gas industry to back hydrogen, including a perceived familiarity with the processes involved (compression, pipeline delivery, combustion) – but hydrogen isn't as similar to natural gas as it might seem.

Hydrogen used for energy is generally classified as 'grey', 'blue' or 'green'. The great majority of hydrogen produced today is 'grey hydrogen', produced by the natural gas industry as a by-product of refining natural gas. Blue hydrogen is captured from natural gas during the process of steam methane reformation. In theory the resulting emissions can be curtailed through carbon capture and storage, at some expense. For reasons I explain later in this chapter, carbon capture at enormous scale is implausible, and blue hydrogen should be understood as the natural gas greenwashing that it is. Green hydrogen is based on the notion of using renewable or nuclear energy to separate hydrogen (H) from water (H_2O). This is the best of the hydrogen ideas, but still has challenges in terms of cost and round-trip efficiency.

Whichever category we're talking about, to use hydrogen in an energy system you need to do the following things:

1. Separate the hydrogen from water or some other molecule.
2. Compress the hydrogen or cryogenically cool it, to make transportation feasible.
3. Store the hydrogen in some safe pressure vessel.
4. Transport the hydrogen to where it is needed.
5. Decompress the hydrogen so that you can utilise it.
6. Either burn the hydrogen by mixing it with oxygen, like a traditional engine (and maybe use that to turn a generator to make electricity), or . . .
7. Put the hydrogen through a 'fuel cell' to turn it directly into electricity.

All these things take energy to achieve. All require machines. All cost money. As you know, I love Sankey diagrams, so let's express hydrogen's handicaps that way. Figure 4.5 looks at the case for hydrogen as a battery for energy storage. From electrons in, to electrons out – what energy nerds call 'round-trip efficiency' – batteries are already around 90% efficient. The best-case round-trip efficiency for hydrogen is only about 41% using fuel cells, or 35% using combustion. To afford to be this wasteful, you would have to double the amount of solar or wind power in the mix.

4.5: The efficiency problems with using hydrogen as a battery. A large amount of energy is wasted during the process.

All electric – energy transformations for storage

100% STARTING ELECTRICITY	Transmission (~5% loss) →	Battery storage (~5% loss) →	Inverter losses (~3% loss) →	88% AVAILABLE FOR USE

12% WASTE

Green hydrogen – combustion – energy transformations for storage

100% STARTING ELECTRICITY	Electrolysis (~17% Loss) →	Compression (~15% loss) →	Transportation (~3% loss) →	Combustion (~50% loss) →	35% AVAILABLE FOR USE

65% WASTE

Green hydrogen – fuel cell – energy transformations for storage

100% STARTING ELECTRICITY	Electrolysis (~17% loss) →	Compression (~15% loss) →	Transportation (~3% loss) →	Fuel cell conversion (~40% loss) →	41% AVAILABLE FOR USE

59% WASTE

What about using hydrogen to power trucks, cars, mining vehicles, aeroplanes, trains or buses? I actually know something about this. I developed hydrogen tanks with the US Department of Energy, a technology we ultimately sold to a consortium that included Toyota, Audi, Chrysler, Porsche, McLaren and others. I don't think there will be no hydrogen in the future, just that there will be less than you may think, and it will be more expensive. Figure 4.6 looks at the case for hydrogen for transportation, again compared to batteries. Batteries probably achieve something like 85% efficiency in applying torque to the wheels. If you use hydrogen, either by combustion in an engine or through a fuel cell, the vehicle will be less than half as efficient, at 34–37%. Yes, there may be some remote applications where hydrogen will be important, but remember that batteries are getting cheaper and better, so the target is moving.

Then there are advocates saying that we need hydrogen for high-heat industrial applications – the places we currently burn natural gas. The

4.6: The efficiency problems with using hydrogen for transport. A large amount of energy is wasted during the process.

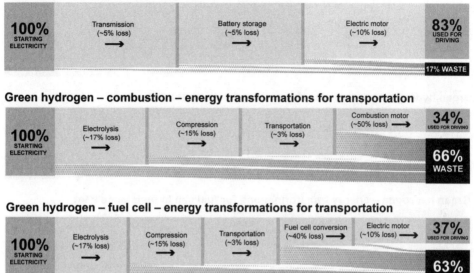

All electric – energy transformations for transportation

| 100% STARTING ELECTRICITY | Transmission (~5% loss) → | Battery storage (~5% loss) → | Electric motor (~10% loss) → | 83% USED FOR DRIVING |

17% WASTE

Green hydrogen – combustion – energy transformations for transportation

| 100% STARTING ELECTRICITY | Electrolysis (~17% loss) → | Compression (~15% loss) → | Transportation (~3% loss) → | Combustion motor (~50% loss) → | 34% USED FOR DRIVING |

66% WASTE

Green hydrogen – fuel cell – energy transformations for transportation

| 100% STARTING ELECTRICITY | Electrolysis (~17% loss) → | Compression (~15% loss) → | Transportation (~3% loss) → | Fuel cell conversion (~40% loss) → | Electric motor (~10% loss) → | 37% USED FOR DRIVING |

63% WASTE

case against that is plain to see in Figure 4.7. Again, if we are thinking about a zero-emissions pathway that starts with renewable (or even nuclear) electricity – which is really the only option we can entertain, given the urgency – electric heating through either resistance or induction, both of which can satisfy the very great majority of industrial processes, provides 88% efficiency, significantly better than hydrogen in combustion, or more than twice as good if we go through electricity to hydrogen to a fuel cell to electricity and back to resistance heating. It soon becomes apparent that all those energy transformations are an efficiency cost, a fundamental fight with the laws of physics that you can never win.

Clutching at straws, a hydrogen advocate might argue we need it for building heat. This is maybe the worst idea of all and should be exposed for the sheer audacity of the (ahem) gaslighting. The idea is that we can put hydrogen into our natural gas pipes, pump it into our homes and use it like gas. Well,

4.7: The efficiency problems with using hydrogen for high-temperature heating applications. A large amount of energy is wasted during the process.

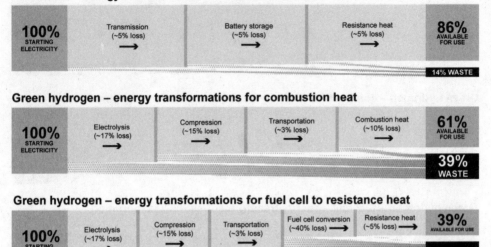

All electric – energy transformations for resistance heat

| 100% STARTING ELECTRICITY | Transmission (~5% loss) → | Battery storage (~5% loss) → | Resistance heat (~5% loss) → | 86% AVAILABLE FOR USE |

14% WASTE

Green hydrogen – energy transformations for combustion heat

| 100% STARTING ELECTRICITY | Electrolysis (~17% loss) → | Compression (~15% loss) → | Transportation (~3% loss) → | Combustion heat (~10% loss) → | 61% AVAILABLE FOR USE |

39% WASTE

Green hydrogen – energy transformations for fuel cell to resistance heat

| 100% STARTING ELECTRICITY | Electrolysis (~17% loss) → | Compression (~15% loss) → | Transportation (~3% loss) → | Fuel cell conversion (~40% loss) → | Resistance heat (~5% loss) → | 39% AVAILABLE FOR USE |

61% WASTE

you can only put about 20% hydrogen in the mix before you need to replace the pipe networks because the metal of the existing pipes doesn't support hydrogen. This would be prohibitively expensive. Figure 4.8 shows that, once again, electrifying everything will be at least twice as efficient.

You might think I've gone out of my way to lampoon hydrogen. I have. There are not infinite financial resources in the world. Investment in hydrogen is not as wise as investment in electrification. There will be some uses for hydrogen – for example, in creating ammonia and other fertilisers without natural gas, a hugely ambitious and worthwhile hydrogen project. And hydrogen may make more sense for particular places: Western Australia, for example, is remote and could be a huge producer of hydrogen. It is one of the only places in the world where it *sort of* makes sense. But let's not let this distract us from focusing on electrification as the best, most efficient pathway to decarbonisation.

4.8: The efficiency problems with using hydrogen for low-temperature heating applications. A large amount of energy is wasted during the process.

All electric – energy transformations for heat pump heat

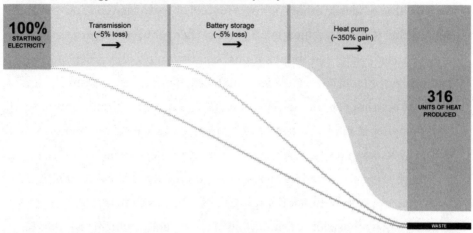

Green hydrogen – energy transformations for fuel cell to heat pump heat

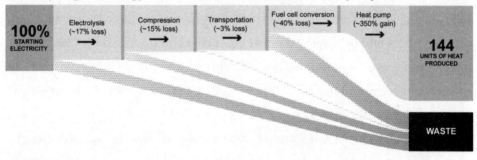

All electric – energy transformations for resistance heat

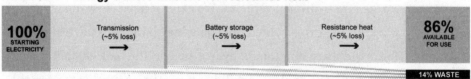

Green hydrogen – energy transformations for combustion heat

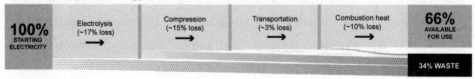

Some of our trading partners, including Japan, have really bought into the hydrogen story. So has Germany. During World War II. Japan and Germany had scarce domestic liquid fuels, and so without diesel or oil were forced either to invade other countries for fuel or try to turn coal into a high-density fuel through gasification. This sent them down the hydrogen path. For countries concerned about national security and energy independence, hydrogen may seem like a great idea: in theory, if one supplier stops sending it to you, another can fill the gap, whereas if your connection to your electrical supplier is cut, you are stranded. Much of the hype about hydrogen is driven by issues of national security and energy independence.

There are also vested industry interests involved. One champion of hydrogen, the International Energy Agency (IEA), isn't in fact an independent agency, but basically a trade group of fossil-fuel-producing nations. Its assessment that hydrogen could provide half the world's electricity by 2050 is laughable, as this would require doubling or tripling the global supply of renewables and deploying them at great cost and inefficiency. The IEA's reports are worth reading, however, if only to see how desperately the fossil-fuel industry would like the future to look like the past, with all the world's energy flowing through its pipes.

One last thing. When I worked with Toyota and the US Department of Energy on natural gas and hydrogen tanks, we owned two natural gas vehicles and worked on hydrogen vehicles. We experienced close-up what it is like to work with hydrogen, including blowing up the tanks to test their 'safety'. Here's the scary thing about hydrogen . . . If the tank explodes, the pressure wave from the explosion will likely collapse your lungs. If that doesn't kill you, the invisible gas will suffocate you. If by some miracle you survive that, the hydrogen will then ignite in an invisible flame, frying you to a crisp. Oddly, the people who want to sell you hydrogen don't tell you these things. I wonder why?

Carbon capture and storage

Without doubt there will be someone out there asking, 'What about carbon capture and storage with existing fossil fuels?' We might as well address that red herring here. It is definitely possible to capture carbon dioxide, and it is possible to store it. But there are major objections to considering this a serious solution. Having delayed reducing emissions for so long, the world has now bet enormously on 'negative emissions' – that is, carbon capture and storage, or CCS. So much so that we are planning on capturing carbon from BECCS (bioenergy with CCS) on a huge scale.

When you burn a hydrocarbon fuel, the carbon atoms get oxidised with two oxygen atoms each, and the fuel is turned from a solid or a liquid into a gas – and gets about 5000 times bigger in the process. Even if you compress it back down into a liquid (which requires a huge amount of energy), the carbon still ends up three times bigger than when it came out of the ground as a fossil fuel. This means that to capture and store it, we would need to fill giant underground reservoirs, larger than the reservoirs it came out of in the first place.

In its most recent report, the Intergovernmental Panel on Climate Change (IPCC) outlined the best-case scenario for CCS. It assumed burying huge amounts – 10 billion tonnes – a year of CO_2. That is as many gigatonnes of stuff as the *entire* fossil-fuel industry currently rips out of the ground. It imagines an industry as large as the entire fossil-fuel industry to transport, process and bury CO_2, and that this industry will somehow magically spring into being by 2050.

Carbon capture and storage is also expensive. Adding CCS to a fossil fuel will inevitably increase the fuel's cost. Anti-tax ideologues across the world, including those in Australia with a 'technology, not taxes' mindset, make it unlikely that there will be tax revenue to pay for all this CCS. Yet for nearly every application, electrification is already the cheapest path, and getting cheaper. Given this, where will the fiscal motivation for CCS at a massive scale be, particularly come mid-century once most things are electrified?

It is very unlikely that CCS for fossil fuels will be anything but a very expensive niche. It is a fraught fantasy of the fossil-fuel industry trying to extend its relevance. We should be enormously sceptical and demand far more transparency about any government plans to engage in such boondoggles.

More than 100%

Australia has the potential to install more electricity generation than we need. Renewable energy and proponents of sustainability grew out of the efficiency movement of the 1970s and 1980s, when people barely imagined getting to 100% renewable, let alone 120%, 200% or 700%.[22] What's more, getting only to 100% isn't the cheapest option, now that the generation costs of wind and solar (three or four cents per kWh and falling) and rooftop solar panels (six or seven cents per kWh) are so low. If, for example, we designed our solar installations for the winter minimum as opposed to the summer peak, we'd need to install 50% to 100% more. But since that would lower the need for seasonal storage technologies, and even for day-to-day and week-to-week storage, it would likely lower the overall cost of electricity.

In a study I did for the US, presented in my book *Electrify*, I showed that installing roughly 125% capacity of wind and solar would eliminate seasonal storage requirements entirely and allow the 25% over-capacity in the summer to be used for generating electrofuels or hydrogen or even just over-producing energy-intensive industrial products on the very cheap electricity of that season.

This idea of abundance still sits uncomfortably within the green community, because it is in such contrast to a history of selling scarcity. Without doubt, getting to 100% renewables and beyond is made easier by setting the target even higher – 125%, 150% or even 700%.

Efficiency

This book is really about encouraging electrification, so I don't talk much about traditional energy efficiency. Efficiency has been a mainstay of the energy conversation since the 1970s. To many people this means that a slightly more efficient natural-gas heater, or a slightly more efficient car, is good enough. But we need to get to zero emissions and we can't efficiency our way to zero, regardless of how 'efficient' those heaters and cars become. Getting to zero requires transformation, and electrification is the path. An electric car uses one third of the energy of a petrol car if the electric car is charged from wind or solar. That's a huge efficiency win that will get us to zero emissions. Heat pumps, which can create three or four times as much heat as the electricity they use to move or pump that heat, are effectively 300–400% efficient heating devices. Compare that to a natural-gas heater's 80–90% efficiency, or a fireplace at 50–60% efficiency. It turns out that electrification is the efficiency we've always been looking for.

That said, Australian houses are some of the worst in the world in their leakiness and poor insulation. Leakiness is all the holes and gaps between doors and window frames that allow cold or hot air to come into the home, requiring more energy to compensate in heating and cooling. Most of our homes have little or no insulation in their wall and roof cavities, and for the most part we don't have double-glazed windows. All these simple building technologies would make the process of electrification and decarbonisation easier and cheaper. But insulating your home or closing the gaps economically is generally only something that happens when the home is newly built or undergoes a significant renovation. We should be promoting better building codes and standards, and they should apply not only to new builds but also to any retrofits.

Long-distance transmission

The wind is probably blowing in South Australia at times when it isn't in Queensland. The evening peak load for the Victorian and New South Wales electricity grids coincides with the peak generation of wind and solar in Western Australia. Tasmanian hydroelectricity and wind power can help all the other states to balance their grids. No individual wind turbine is always spinning, but many turbines distributed over different geographic regions will produce wind most, if not all, of the time. On a continent as large as Australia, the three-hour time difference between the east and west coasts means that midday solar generated in the east can help with the morning peak loads in the west, and late afternoon sunshine in the west can help the east later in the day. To achieve this, we need long-distance transmission of electricity.

Long-distance transmission sounds expensive but is actually quite cheap, especially considering the advantages it confers in robustness and reliability. It sounds expensive because the capital cost typically runs to about a million dollars for each kilometre of transmission, delivering 1 GW of power. But transmission lines last a long time, more than 25 years. A little maths reveals that at this cost, transmission lines over 500 kilometres will increase the cost of electricity by less than two cents.[23] That's a tiny fraction of the generation cost. This topic is being extensively studied globally. There is currently no plan to connect Western Australia to the rest of the network – but if I had to predict the future, given the tripling or more of electricity that will need to be generated and delivered for Australia to become a renewable energy superpower, I reckon it will happen.

Ammonia, electro-fuels, compressed air and other odd ducks

There are a few other long-odds options that might make a meaningful contribution to the Australian clean-energy transition. There are other potential fuels, such as ammonia and other 'electro-fuels', all of which have some advantage or disadvantage relative to hydrogen, but none of which yet looks to have the advantages of plain old diesel. Compressed air is often touted as an energy storage mechanism, which is in principle quite simple – compress the air when you have energy and allow it to spin a turbine when you decompress it. The problem with compressed air is that as much as 50% of the energy is lost. It's not a very efficient battery.

The broad brushstrokes of the energy transition are all well outlined at this point and easy to understand. The market will choose the balance of the mix, and at this point it is highly unlikely that something not mentioned here is going to change the game or the outlook. Hardware technologies typically take a decade or two to go from idea to first deployment. If the idea isn't already out there and getting traction, it's unlikely to be a major contributor to our progress towards zero emissions.

Chapter 5
Electrify (almost) everything!

- Electrified appliances and vehicles are our secret weapons in the fight against climate change.

- Thermodynamics tells us that we can run the same country we do today using less than half the energy, just by electrifying all our machines.

If you only read one chapter in this book, this is the one I'd have you read. The future is electric. Increasingly, academia is reaching this conclusion, fundamentally because physics and thermodynamics show that it is a good idea.[24] Australia needs to hear this message loud and clear lest we be delayed by the promises of false solutions. Our energy-related emissions require us to make an enormous commitment to electrification as the national strategy, and to make it now.

Thermodynamics and high-school physics

You may remember your high-school physics teacher muttering something about the 'laws of physics'. The most famous of those laws are the laws of thermodynamics. There are three big ones that are worth remembering.

1. You can't win; you can only break even (meaning no free energy).

2. You can only break even at absolute zero (it's hard to convert between energy types efficiently).

3. You can't reach absolute zero (even getting close to winning is impossible).

Of course, these are informal translations of the laws. The point is, converting energy from one source to another is a losing proposition because it is inefficient. The first law says your Uncle Jack can't build a perpetual motion machine out of baling wire and tinnies that will keep the country running on 'free energy'. Whenever you are converting energy from one form to another, you are flirting with these laws. When we burn fossil fuels to create electricity, we are converting chemical energy into heat, the heat into motion, and the motion into electricity. That's a lot of conversions and each conversion is imperfect, meaning we lose some energy each time. These laws, especially the second, explain why electricity is a fundamentally better type of energy to do the things we need to do.

When we burn fossil fuels, we run up against the second law, and more specifically a principle known as Carnot efficiency. Sadi Carnot, considered the father of thermodynamics, was a French physicist who improved the efficiency of steam engines. He demonstrated that efficiency was determined by the difference in temperature between the hot and cold sides of the engine. Even if you don't lose any energy to friction or other mechanical factors, you can't get all the energy out of the fuels. In practice, this translates to hard limits on the efficiency of engines. A small car engine might be 20–25% efficient, a truck maybe 30%. A coal-fired power plant generating steam struggles to get above 40% and very often operates at only 25%.

When electricity is harnessed from the wind, we don't have these combustion losses and 95% of the energy generated can be used to fill batteries, run motors, power lights or heat water and air. Starting with wind or solar and using that to generate electricity is the secret to saving huge amounts of energy. The Australian economy can run on far less energy if we embrace the electrification of everything. In this chapter, I'll outline what this might look like for a range of technologies, from cars to cooktops to the wider national economy.

Electrifying our vehicles

Electric vehicles are approximately three and a half times more efficient in converting energy into motion than their internal combustion engine (ICE) counterparts. This is because there is no engine throwing away 80% or so of the energy in the fuel and converting it into heat. That heat loss is why you can throw a pot under the bonnet of a Land Rover and cook a stew as you drive – there is a ton of wasted energy.

The battery in an electric vehicle weighs more, but the motor is around 95% efficient, and the motor is lighter than the ICE engine. Most electric cars only have one 'gear' because the motor has high torque at all speeds. This eliminates the need for a transmission, which is another source of waste (and weight) with an ICE. Finally, with an electric car you can do

5.1: Most cars today waste 75% of the energy in petrol. We could use heroic efforts to make them a bit more efficient – or we could electrify the vehicle, use energy from wind or solar, and reduce our energy needs to a third or less.

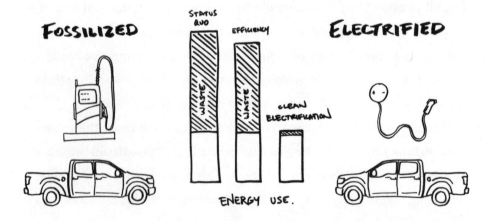

THE BIG SWITCH

regenerative braking, which captures quite a lot of the energy that would otherwise be warming your brake disks, and recharges the battery.

This situation is most clearly described in Figure 5.1, which captures the trend regardless of the vehicle type. Some people will argue until they are blue in the face that big or long-distance trucks will never be electrified, but they will be. Truck drivers can't safely drive for more than 12 hours a day. That's two six-hour stints of 600 kilometres each if they drive at the speed limit. That's already achievable using today's batteries, and you have to remember that batteries are getting better and cheaper. We will likely build dedicated charging infrastructure for these vehicles.

Electrification of space heating

There are four main types of space heating: natural gas, electric-resistance heating (bar heaters), wood fires and electric reverse-cycle air-conditioners (heat pumps). Natural gas heating has an efficiency of approximately 0.9, meaning each energy unit of natural gas is converted to

5.2: Conventional space heating using gas heaters, bar heaters, fireplaces, etc. can be made more efficient by small amounts, but electric heat pumps (reverse-cycle air-conditioners) are already about three to four times as efficient.

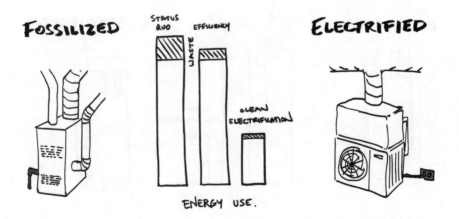

0.9 units of heat. Electric resistance heating has a slightly better efficiency of approximately 0.95. Wood fires have an efficiency of approximately 0.75, converting one unit of energy in a log to 0.75 units of heat in a room. Finally, reverse-cycle air-conditioners have an approximate efficiency of a whopping 3.8 in the average Australian climate. This gives us an odd comparative graph of efficiencies, as shown in Figure 5.2. An existing gas heater is 90% efficient and might be made slightly more so, but a heat pump uses less than one third of the energy to harness the same amount of heat and deliver it to your building.

Electrification of water heating

Hot showers are important to just about everyone at this point, and many of us don't mind warm water for scrubbing our germ-exposed hands either. Just as with space heating, heating water for these uses is 'low-temperature heat', a fairly unscientific way of describing water below boiling point and amenable to being supplied by a heat pump. The same efficiencies apply as

5.3: Conventional water-heating with gas or electric resistance can be made more efficient by small amounts, but electric heat-pump water heaters are already around three to four times as efficient with the same heating result.

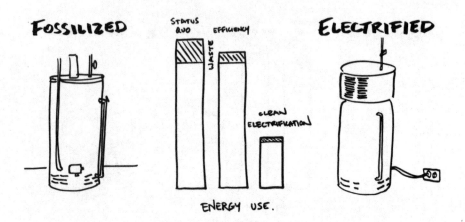

THE BIG SWITCH

for space heating. The winner, once again, is heat pumps, which in effect are 300–400% 'efficient', meaning they use a third to a quarter of the energy to heat the same amount of water. I insulated a hot-tub in my backyard in Wollongong and heat it with a heat pump: when my solar is making more energy than I can use, I dump it into the tub for a relaxing after-dinner soak.

Some people might point out that you can use solar to heat water directly in Australia. The Solahart brand has been a favourite technology for years. It is about 80% efficient in converting sunlight into hot water, which, because solar is 20% efficient at producing electricity and heat pumps are 400% efficient at making heat from that electricity, is about the same as using solar panels and a heat pump. I just know some grumpy old engineers are going to contact me about this, and I look forward to it.

Electrification of cooking

Cooking with gas is much less efficient than cooking with electricity. A gas stovetop has an approximate efficiency of 0.3, and an electric resistive stovetop

5.4: In heating a pot of water with a natural-gas burner, 70% or so of the energy is lost as heat. This is far less for electric resistance stovetops, and even less again for electric induction stovetops.

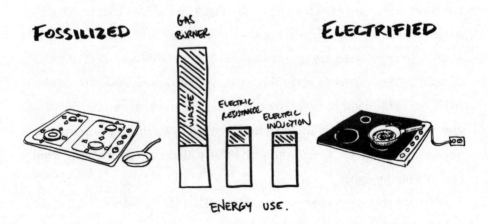

FOSSILIZED   GAS BURNER   ELECTRIFIED

WASTE

ELECTRIC RESISTANCE   ELECTRIC INDUCTION

ENERGY USE.

has an efficiency of 0.7, or even higher for electric induction stovetops. The efficiency of heating by various means is illustrated in Figure 5.4. When you heat a pot of water with a natural-gas burner, 90% or more of the energy in the natural gas gets converted to heat, but 70% or so of that energy is lost because it heats the kitchen or room, not the water. You can feel this energy in the kitchen if you have been cooking for a while. You can feel the heat escaping from underneath the pot and under the sides. Electric resistance is better, because the heat is more directly transferred to the pot, and is often 70% efficient, twice as good as gas. Induction burners are the miracle new technology using the mysterious powers of magnetic fields to heat the pot, not the room. These are as high as 90% efficient at turning electricity into boiling water and in every way are a better cooking experience than cooking with gas – faster to heat, more control, easier to clean, cooler kitchen, lower burn risks for children, cleaner air for the whole family.

Making electricity, not heat

Today we make most of our electricity with coal or natural gas. This is the source of the largest energy losses in the Australian energy system. You can thank that damned second law, and the fact that most of the energy escapes as heat and isn't converted to electricity. Continuing our cartoon guide to the thermodynamics of electrification, Figure 5.5 shows that making all our electricity with wind, hydro and solar would eliminate the 60–70% of waste energy involved in generating electricity the silly way we do it today by burning rocks. The clouds in the cartoon on the left represent the steam rising out of the cooling towers from the coal or natural-gas electricity generation plant – the clouds are the wasted heat escaping into the atmosphere, not making electricity. There is no such waste with the solar panels and wind turbines on the right.

Perhaps you are seeing the trend here. Despite decades of being told to be more efficient, what we should have been told is to be more electric,

5.5: Energy waste compared, before and after electrification.

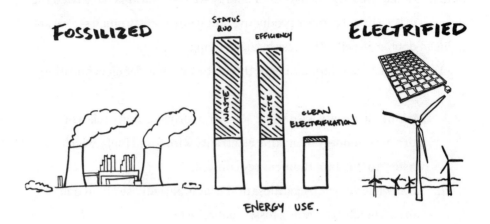

and to make our electricity more renewable, because that is where the real efficiency is – in electrification.

Electrifying industry

People think industry is going to be so hard to electrify, but it needn't be. Industry has a lot in common with the examples we've just discussed. Industry moves things around, which will be more efficient when electrified. It heats a lot of stuff up, to low temperatures and to high temperatures, both of which would be more efficiently done using electric heat pumps and induction heating. Electrochemistry is replacing a lot of traditional industrial processes, including in making steel and other metals. This is lowering the energy used in these intensive industries and making industry more productive.

Electrifying the Australian economy

As simple as these examples are, they capture the essence of what there is to win with electrification: a much more effective economy. Economists measure this and call it energy productivity – a measure of the economic benefit we receive from each unit of energy we use. Nations have tried for decades to get a little bit more productive, but we can pretty much see we are going to double our effectiveness by electrifying.

You can use these simple rules of thumb to advocate for electrification:

1. Making electricity with wind, solar or hydroelectricity takes one third of the energy of making electricity with fossil fuels, which waste two thirds of their energy content.
2. An electric vehicle, regardless of size or type, will use about one third as much energy as a fossil-fuel vehicle.
3. For low-temperature heat like domestic hot water and space heating, a heat pump needs only one third to one quarter of the energy of heating the same thing with fossil fuels.
4. For high-temperature heat, induction heating needs only half to three quarters of the energy that would be required using fossil fuels.

We can do a rough before and after comparison of what it means to commit to an electrified Australia. In Figure 5.6, we fastidiously went line by line through the Australian energy data and used the approximate gains in efficiency by electrifying the majority of everything. Even allowing for growth, and without doing everything perfectly, including some losses in electrical storage and transmission, we can say with some confidence that a commitment to electrification of the Australian economy will more than halve our energy consumption. Traditional card-carrying greenies bearing battle scars from the energy-efficiency wars of the 1970s, '80s and '90s won't like the one big conclusion you can draw from this. We can have the same cars as we do now, just as big, only electric. We can still have the biggest houses in the world, only electric. We can still have industry and business, only electric. We can have what we have now – using only half the energy.

5.6: The difference in energy use (and waste) in Australia's domestic economy if we electrify everything. Less than half the energy use for the same result.

Current domestic energy end-use and waste (2018–19)

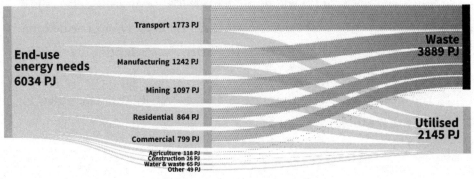

Future electrified domestic energy end-use and waste

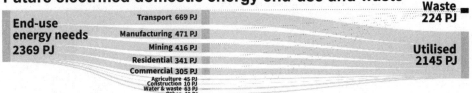

Source: Australian Energy Statistics 2020 and Rewiring Australia

Of course we should also insulate our homes and seal the cracks and ride bikes and use public transport. All these things would lower our energy requirements a whole lot too. The point is, we don't have to be perfect to solve climate change. We just need to be electric.

With electrification, more is better

It is important to remember this about electrification: the more, the better. The more vehicles that are electrified, the easier it will be to find a charging station. The more electric cars exist, the easier it will be to use them as batteries on wheels to absorb our abundant renewable energy. The more homes with electrified heat, the more opportunities for demand response and storage. The more industries that electrify, the more electricity we'll produce, and the easier it will become to electrify everything else. More rooftop solar begets more electric vehicles begets more electric heating systems begets more commercial and industrial electrification, more wind turbines and more batteries.

You see the trend here. The more things we electrify, the easier it gets to electrify all the things.

Chapter 6
Cheap and getting cheaper

- Renewable energy is already cheaper than fossil fuels in many cases.

- Increasing production rates further cuts the cost of a technology. Making enough renewable energy, batteries and electric vehicles to address climate change will more than halve the cost of them.

- If we keep growing at the current growth rate, we could fully decarbonise the planet by 2040.

We have the technology to create a carbon-free future, but can we afford to make the switch? It may seem sacrilegious to discuss costs when considering the future of our planet, our species, and the beautiful critters and plants we share Earth with. It's dismal to have to justify the economic cost of doing the things that will make our future better. But in this chapter I sharpen my pencil and show you how, in fact, saving the planet will save everyone money.

In his excellent book *The New Climate War*, Michael Mann plainly lays out the delay tactics being used by the fossil-fuel industry. Sowing confusion, slowing things down with bureaucracy and selling unlikely miracle technologies are how the industry will try to hold on to its profits, at your children's peril. The Australian government is a perfect example of this. The brilliantly simple phrase 'technology, not taxes' sounds like what we would all like: a technological wonder to save the day. But this masks the urgency of the actions required and the fact that not only are we not currently taxing fossil fuels – we are subsidising them.

It is a government's job to lead its people into a future that's better than the present for its citizens. This is not a job that requires sitting on one's arse, waiting for the solution to happen. It demands setting policy based on the best advice and data available. One has to have an opinion about the future to lead – a vision. That's why people love John F. Kennedy's 'We choose to go to the moon' speech. It set an agenda that a whole nation, in fact the whole world, could get behind.

As explained in the last chapter, it is abundantly clear that electrification is going to do most of the heavy lifting in addressing climate change. This has been obvious to many for a decade now, but it still eludes governments on both sides of Australian politics. No one has dared set a national electrification agenda, yet that is what we need. To wait for a newer or cheaper technology to come along is to delay too long.

We already know (and a hundred years of data supports this) that when it comes to technology, much of the cost reduction comes from learning by doing. The more we use a new technology, the cheaper it gets. As we electrify and decarbonise, Australia will surf this wave of cost reduction. Technological improvements in the last two decades have already dropped the cost of critical technologies – solar, wind and batteries – to below that of fossil fuels. The scale of decarbonising the world's energy supply is sufficient to drop the cost of renewables by half, such that they will trounce the cost of fossil fuels. In this chapter, I'll explain why.

Electricity is cheap, and getting cheaper

Already, generating clean electricity is extremely cheap and getting cheaper, and some of it will become cheaper still – provided we don't screw up with

6.1: Lazard's levelised cost of energy comparison, showing wind and solar are the cheapest source of new generation.

Levelized cost of energy comparison – unsubsidized analysis $USD

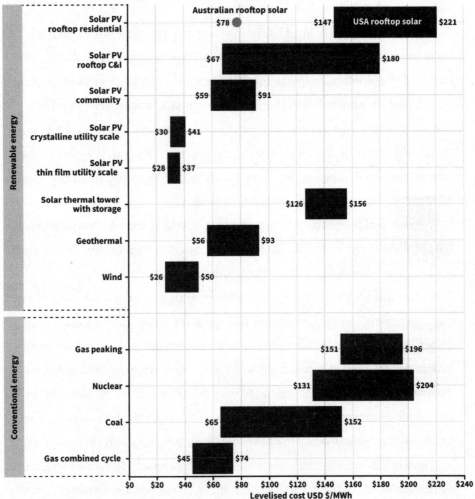

the wrong rules and regulations. When energy nerds compare the prices of different types of energy, they talk about the levelised cost of energy (LCOE). This is how much a particular technology costs per kilowatt hour (kWh) when all its lifetime costs are taken into account (such as the costs of building, operating and decommissioning a plant). The asset management firm Lazard, which tracks LCOE to guide investments, has data showing how much cheaper renewable energy sources are compared to fossil fuels. Its latest report places utility-scale solar at about 3.7 cents per kWh and wind power at about 4.1 cents per kWh, while natural gas clocks in at around 5.6 and coal at 10.9.

These impressively low LCOE numbers refer to utility-scale installations. Rooftop solar can be even cheaper. If you're generating electricity yourself, you don't have to pay for distribution. Australia has lowered the cost of rooftop generation so much that 'behind the meter' energy – that is, energy generated on our rooftops, without relying on a utility – is cheaper than the cost of distribution alone from a centralised plant. We can't make all the energy we'll need in the future this way, but we can make an awful lot of it, and it is already cheap and getting cheaper.

A friend and fellow Aussie expat, Andrew 'Birchy' Birch, wrote an influential piece about replicating the Australian model of rooftop solar in the US. He showed how most of the costs in the US are 'soft costs', or those not directly tied to hardware. These include permits inspection, overheads, transaction costs and sales. The Department of Energy agrees with him and its current $1 per watt target is focused on eliminating soft costs. Australians don't tolerate heavy-handed bureaucracy well and hence we have succeeded in eliminating soft costs in solar. We need to take the same attitude to heat pumps, cars, vehicle-charging infrastructure and household batteries to ensure that future savings are passed on to households, not nabbed by bureaucracy.

Here is the transformative point about rooftop solar. Because there are no transmission and distribution costs, it can be phenomenally cheap. Even if utility-scale generation were free, we don't know how to transmit it and distribute it to you and sell it to you for less than the cost of rooftop solar. Transmission, distribution and billing costs are often more than half the cost of Australian electricity and add up to as much as 15 cents per kWh. This doesn't mean the whole world will run on solar – but if we are looking to make the lowest-cost energy system, an awful lot of our energy will come from our rooftops and our communities.

Renewables are going to get even cheaper

Wind and solar are getting cheap so quickly that it's even hard for innovators to keep up. I started Makani Power, a kite-powered wind energy company, in 2006. The idea was to produce wind energy at three to four cents per kWh, cheaper than natural gas and five to six times cheaper than other wind-powered electricity at the time. The project was truly awesome, building wings the size of 747s, tethered by a giant cable, that flew in circles at 200 miles per hour, undergoing eight Gs of acceleration while producing megawatts of electricity. With investments from Google the company followed an exciting development trajectory, culminating in an offshore deployment and demonstration in Norway in partnership with Shell.

In the meantime, however, the wind industry at large also made historic strides, and is now routinely deploying turbines at four to five cents per kWh. In 2020, Makani shut down due to this evaporated advantage. The technology and execution were sound, but the industry found its own way to slash costs, just by the improvements that come deploying at massive scale. Although Makani's technology didn't win the cost battle, it was part of an enormous movement and ecosystem of global innovators responsible for driving down costs and making wind, solar and batteries competitive with fossil fuels.

In 2011 I started another company, Sunfolding, with Leila Madrone and Jim McBride. We initially focused on building tracking devices – machines that make sure the solar technology follows the path of the sun through the sky accurately. But the relentless march of photovoltaics' price improvements beat us out of that game too, and we 'pivoted' – as they say annoyingly in the Silicon Valley – to tracking devices for PV. We are still in the game and are now selling our technology into industrial solar plants at basement-level prices that come out at around two cents per kWh – lower than we ever imagined, and far lower than any fossil-generated electricity.

6.2: Learning curve of Ford's Model T.

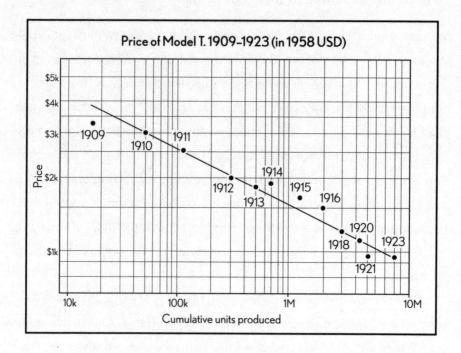

Price of Model T. 1909–1923 (in 1958 USD)

Source: Abernathy and Wayne, 'Limits of the learning curve', *Harvard Business Review*, 1974.

There are two ways to reduce the cost of energy. One is by inventing better mousetraps; the other is by producing mousetraps in gobsmacking quantities. The first, called 'learning by researching', is typically measured by cumulative investment. The second, 'learning by doing', is measured by cumulative total production. Makani was about building an entirely better mousetrap, but it couldn't make mousetraps in quantity. Sunfolding was one of many small component improvements. It was an invention, but it wasn't the whole mousetrap; it was like a better mousetrap spring. Sunfolding's tracking technology is good for taking five to ten cents out of the roughly $1 per watt. Half of this cost saving was in the hardware we invented, but crucially half was in reducing installation labour costs. It is these small efficiencies in materials and labour that typify the learning-by-doing cost savings. As these examples illustrate, and as empirical studies have shown,

6.3: Learning curve of photovoltaic module price.

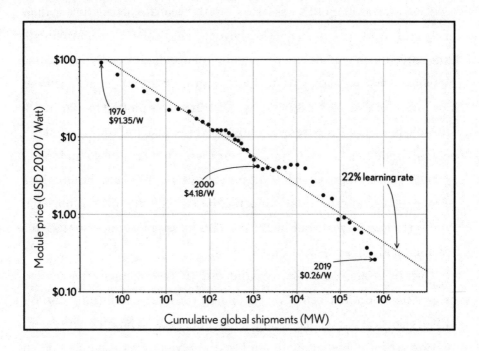

Source: Haegel et al., 'Terawatt-scale photovoltaics: trajectories and challenges', *Science*, 2017.

we must invest heavily in both these capacities to maximise long-term cost reductions in zero-carbon energy.

It is learning-by-doing that gets the job done most predictably. As we've seen, the solar and wind industries are improving, getting cheaper and cheaper with every generation of innovation. Learning-by-doing improvements are measured in 'learning rates', defined as the percentage by which the price falls after investment in a technology has doubled. One of the first observations of these learning rates is known as Wright's law, governing the cost of aeroplanes and the equivalent for automobiles – for example, tracking the decrease in the price of Ford's Model T as production increased, as shown in Figure 6.2. Moore's law, the jaw-dropping exponential increase in integrated circuit density, can be viewed as a version of this same idea.[25]

In the case of electricity generation, solar PV is learning at a rate of about 22% and wind at about 12% – as fast as or faster than fossil fuels during their early 20th-century cost-reduction heyday.[26] For solar, this approximate 20% reduction in module cost per doubling of installed capacity has become known as Swanson's law, after Richard Swanson, the founder of SunPower Corporation. The progress made by this learning is plotted in Figure 6.3, showing how, despite extreme economic events (like the 2008 recession), solar PV modules have continued their march towards lower and lower costs.[27] Not only that, but in just the past five years, global new installations of renewable energy have outnumbered the installations of fossil-powered energy (by nearly two to one in 2018).[28] This means more opportunities for learning, and more opportunities to reduce costs.

Currently, about 250 GW of wind and 125 GW of solar are installed around the world. To electrify everything, we will need about 10,000–20,000 GW of electrical power. (The exact number depends on how the world population grows and what quality of life is enjoyed by what percentage of humans.) That means the cumulative production of solar panels and wind turbines still needs to double in scale many times to reach the capacity we need. If costs are falling at 20% with each doubling, after three doublings the cost will be about 51% of where it started; after four doublings 41%; and after five doublings, only 33%. Given the scale of growth required there is ample opportunity to bring costs down even further, making renewables even cheaper than their fossil competition.

Pause on that thought for a moment. If we commit to wind and solar at sufficient scale to address climate change, that commitment alone will likely more than halve the cost of renewables, yet again – a nail in the coffin of fossil fuels. Electricity will finally (well, almost) be 'too cheap to meter' – as they used to say about nuclear power.

All of this represents a rare opportunity for industry, small and large. The myths of Silicon Valley hold that disruption is always good and that progress is made by unconventional founders turning the world on its head. That model has worked in software, but in hardware, especially in infrastructure, it doesn't really work. These fields are naturally conservative, due to the graver consequences of failure and the need to guarantee machines that work reliably for 20 years or more. As we've seen, progress is predictably achieved through consistent investments in research, coupled with manufacturing at massive scale. We do need start-ups to innovate, and we even need crazy breakthrough ideas, if only because they inspire us to think bigger – but what we need most is large companies to seize on these innovations and scale them up, and for governments to commit to an energy strategy that enables those companies to make appropriate long-term capital investments. We need a national strategy that replaces 'technology, not taxes' with 'commitment and capital'. When we commit, the capital will go there, and the costs will come down and the electrified future will be the low-cost abundant energy future we have always been hoping for.

It's already underway

If we take these growth rates of the critical technologies, we can ask ourselves two simple questions. (1) How much cheaper will it get? And (2) when will this cheapness get us to zero emissions? The International Renewable Energy Agency (IRENA) keeps statistics on the installed base of global renewables. Looking over the past ten years, we see growth rates in hydroelectricity that average 2.5%, marine energy at 0.5%, wind 12.5%, solar 22.2%, bioenergy at 6.1% and geothermal at 3.6%.

Those are pretty astonishing numbers. If we merely keep increasing our production rates of wind, solar, hydro, bioenergy and the other renewables, they will be supplying *all of the world's* energy before 2040. Of course this is achieved through the magic of exponential growth but it does tell you we are on an inevitable path of electrification. It tells us that we can do it, and that we are already growing the critical industries nearly quickly enough.

6.4: Installed renewable energy over the past ten years, modelled forward at current growth rates through 2040.

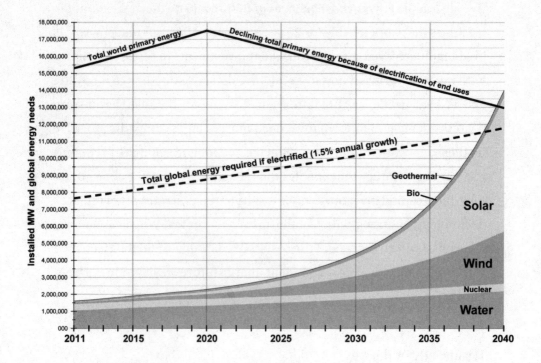

Not only that, but because of the learning curves we have just discussed, the 30 times increase in solar production will lead to cost reductions of 80%. In the future, solar will cost 20% of what it does today – and it is already cheap. For wind it will mean cost reductions of 65%. Batteries are dropping in price even more quickly. To those who tell you it can't be done, you can respond quite simply: it is already being done, and we could make it go even faster.

Chapter 7
Electrifying our castles

- Household emissions make up the largest chunk of emissions in our domestic economy.

- Electrifying our homes will slash our emissions while saving households money on energy bills.

- Electrifying our houses now can generate over $300 billion in household savings by 2035.

Australians love their castles. Generations of economists have hyperventilated about our 'over-investment' in property, and millions of metres of newspaper columns have fretted over whether we have too much debt, not enough debt, too many renters, not enough renters, and whether we should live in more high-rise or less, but it doesn't seem to alter our desire to put down roots and furnish the result with a pizza oven, coffee maker and giant televisions. We build the biggest homes in the world. It is unclear whether we should be proud. We apparently all 'need' air-conditioning and other luxuries of the modern world, including massive garages and a media room.

In those garages, on the street outside and even up on cinder blocks in our front yards are Australia's other obsession: our cars, utes, trucks, vans and four-wheel drives. Similarly huge amounts of newspaper space are dedicated to which car is best, whether best value, highest performance, most nostalgic or most bogan. They are growing in size, just like our houses, and they are all starting to look strangely similar.

Of the decisions we make around our kitchen tables, the daily ones we make about our castles and cars have the greatest impact on the climate. Decades of environmental advice to reduce, reuse and recycle have only had a tiny influence. The key thing to understand is that a small number of infrequent decisions determine what your actual climate impact is. These decisions are about what is on your roof (solar or not), what is in your garage (electric vehicles or not), what is in your kitchen (electric cooking or not), what is in your basement or alongside your house (electric heating and electric hot water or not), and whether you have a battery (or not). These are things you buy once every ten years, not every day, and once you have bought them they become the infrastructure of your life. If you make the right choices (electrification), they lock in climate-friendly behaviour, every day. If you haven't made these big decisions, then the only climate impact you can have is small, and frustratingly spread across thousands of small purchasing decisions. Which of these cans of tinned tuna is better for the planet?

Wise governments should be meeting the voters in their homes with climate and energy policy that helps us to improve our castles and cars. Focus on the infrastructure of our daily lives as we help with every household's journey to electrification and getting to zero emissions. It needs to be a national project, not a culture war. The technologies are basically ready now, and this transition can be extremely good for Australians. We get more bang for our emissions buck with early reductions, and we could lead the world in residential electrification, which is why we should do this proactively as a nation, starting yesterday. Also, it'll save you money, your kids will be healthier, and the air and water in your community will be cleaner.

Household emissions

Households represent the biggest chunk of emissions in our domestic economy: 42%. We can see the breakdown in Figure 7.1, where I rearrange the United Nations Framework Convention on Climate Change (UNFCCC) emissions categories into the Australian sectors that are responsible for them, including our domestic emissions, as discussed earlier. We then break down these emissions into household activities. What may be surprising to some is that our vehicles are responsible for the largest portion of household emissions, closely followed by electricity use, since most large Australian electricity grids are still powered primarily with coal and gas. Overall, energy emissions are about 79% of household emissions, followed by agriculture (what we eat) at 17%. Potentially surprising is how little solid waste (rubbish) contributes to our emissions. Waste reduction is an area frequently championed by large corporations – perhaps because they can be confident that no one likes rubbish, so it's an easier way to market their green credentials than reforming their supply chains. Seeing our household emissions laid out like this makes it abundantly clear that we need to decarbonise our electricity generation, our transport and our gas appliances – all with urgency.

I'm going to focus on the home because it is tangible, but our commercial sector (our small businesses, retail and commercial buildings) have very similar energy-use patterns and machinery. The recipe for decarbonising our castles also applies to most of the commercial sector, which means the low-hanging fruit of another 29% of our domestic emissions, or 71% in total.

7.1: Breakdown of Australia's domestic emissions after being separated from export and trade emissions. Our households contribute the largest chunk of our domestic emissions, followed by our businesses, which have similar electricity, space heating and water heating needs. Electrifying the devices in our homes and businesses can significantly reduce our emissions.

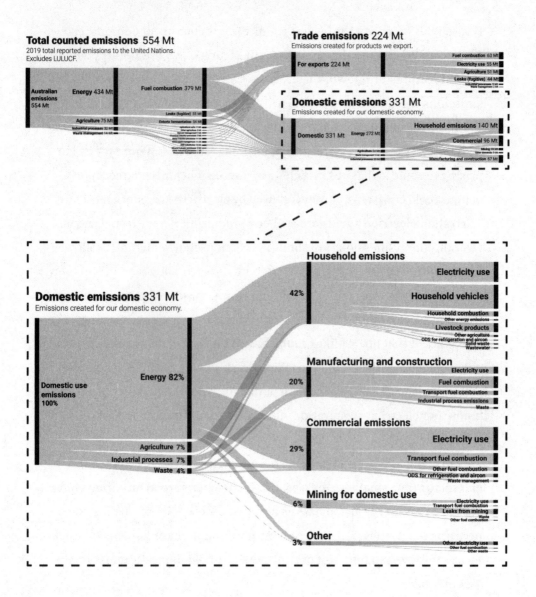

7.2: Average Australian household emissions breakdown. Energy use is by far the largest emitter, led by vehicles and followed by electricity from our largely fossil-fuel-based electricity grid.

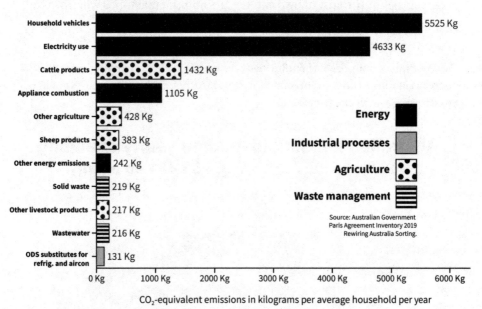

CO₂-equivalent emissions in kilograms per average household per year

Source: Australian Government Paris Agreement Inventory 2019 and Rewiring Australia.

Household spending

The average Australian home spends about $5000 a year on energy use, which includes about $3000 on vehicle fuels, $600 on gas and $1600 on electricity, with small amounts spent on wood, too. This is shown in Figure 7.3. Keep in mind that this is the mythical 'average' house, with about 2.6 people and 1.8 cars, so a house with five people and three cars could easily spend far more than this – as they say, 'your mileage may vary'. Five thousand dollars a year is a significant chunk of money, and it's even more significant for lower-income householders (the bottom 20%), who spend nearly twice as much of their pay cheque on energy as the top income quintile (the top 20%). This can be seen in Figure 7.4. These are costs which, with the right government action spurring the right public–private partnerships and investments, could be practically eliminated over the next three to five years with the savings from home electrification. This might sound like an exaggeration, but it is not. The cost of energy will fall, lowering the cost to operate your house and your car(s), but the cost of owning the cars and the machines will drop too, making the savings bigger than your total spend on petrol, gas and electricity

7.3: Australian household spending per year for the middle-spending quintile. We spend a significant amount of money each year on energy but these costs could be cut drastically with electrification.

Australian household median yearly spending
~$73,000 total spending, ~$5,451 on energy

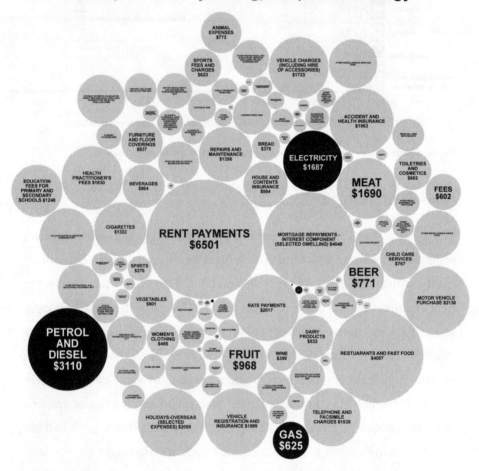

Source: Australian Bureau of Statistics Household Expenditure Survey 2015–16, middle expenditure quintile. Adjusted with ABS Consumer Price Index for June 2021.

today. Perhaps the biggest incentive for Australians to take action on climate change is the hip-pocket savings that are waiting for us if we do. We have so much sunshine, and renewable technology has dropped in price so much, that climate action has become an economically advantageous move.

7.4: Weekly household spending by quintile. Lower-income homes spend a higher proportion of their pay cheques (nearly double) on energy costs.

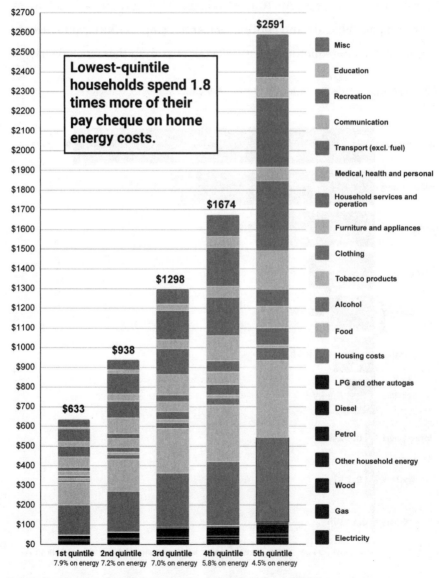

Source: Rewiring Australia, ABS Household Expenditure Survey and ABS CPI.

In the rest of this chapter I'm going to explore how we currently use energy, how that will change with electrification, and the consequences for household economics.

Conventional household energy use

Energy use in an average Australian household breaks down into a few categories: space heating and cooling, water heating, cooking, vehicles and 'other' appliances (which are mostly electric already) – these include fridges,

7.5: Current average home-appliance energy use, excluding vehicles.

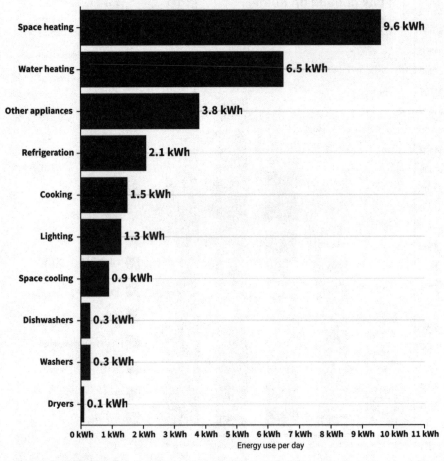

Source: Rewiring Australia

dishwashers, washing machines, dryers, computers, phone chargers and lawn mowers. The majority of total energy use in current households goes to fuelling cars, which consume about 69% of the energy in an average home. Space heating is the second-largest consumer of energy, with 11% of total energy use. Water heating, with 8% of total energy use, and cooking, with 2%, are the other main contributors. The relative portions of all this energy use can be seen in Figure 7.5.

Electrify our castles!

7.6: What an electrified household looks like.

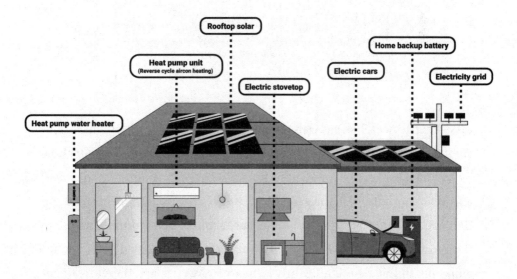

So much of what we hear about climate change is about sacrifice: driving less, taking shorter showers and so on. Maybe we paint this picture for ourselves because of our inherent tendency to be scared of what we have to lose. Maybe the fossil-fuel companies have successfully persuaded us that a future without them will be less joyful. But it is now clear that an electrified future will be a far more abundant and comfortable future for all of us. An electrified household – in all its climate-saving, money-saving glory – is easier to imagine than some might think. It simply replaces any existing

7.7: Average household energy use comparison, conventional household versus electrified household. Electrified households can use less than half the energy for the same result.

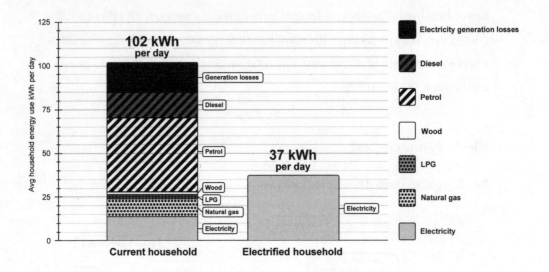

fossil-fuel devices, like gas water heaters and petrol cars, with modern, efficient electric versions. It makes the most of Australia's abundant sunshine by having solar on the roof (something 30% of Australians already have) and a back-up battery in the garage to store the cheap energy being made on the roof. Modern electrified appliances are significantly more efficient than their fossil-fuel counterparts, so not only does an electrified home reduce emissions, but it also significantly reduces energy use and therefore ongoing energy costs. Figure 7.7 shows the difference in energy use between the average conventional home and a fully electrified home.

I hope this feels astonishing to you; it certainly does to me. When we compare the final energy use of a conventional home to a renewably electrified home, the efficiency benefits of electrification become abundantly clear. With the same conveniences, size, warmth and vehicles as a currently fossil-fuelled home, electrifying the average Australian home would cut total energy use by more than half! From 102 kWh per day today, to only 37 kWh

7.8: Average household energy use comparison, conventional household versus electrified household. Electrified households can use less than half the energy for the same result.

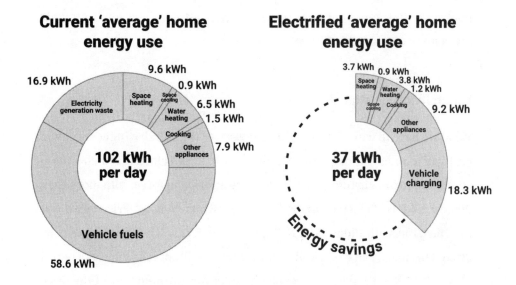

Current 'average' home energy use

9.6 kWh

16.9 kWh

0.9 kWh

Space heating

Space cooling

Electricity generation waste

6.5 kWh

Water heating

1.5 kWh

Cooking

102 kWh per day

Other appliances

7.9 kWh

Vehicle fuels

58.6 kWh

Electrified 'average' home energy use

3.7 kWh   0.9 kWh

3.8 kWh

Space heating

Water heating

1.2 kWh

Space cooling

Cooking

9.2 kWh

Other appliances

37 kWh per day

Vehicle charging

18.3 kWh

Energy savings

Source: Rewiring Australia

tomorrow. This is an enormous win and shouldn't surprise you if you read chapter 5. If we also seal the air gaps in our notoriously inefficient homes (you know there is a draft coming in from somewhere . . .), if we insulate our walls, ceilings and windows, the cost will drop even further. Half the energy for the same comfort, and the majority of emissions gone overnight, all at lower weekly costs. Of course there are numerous other benefits, like safer indoor air quality, not needing to visit a petrol station each week, a battery to get us through blackouts and more.

This may seem too good to be true, or counterintuitive, but remember what we learned in chapter 5 about electrification: electric vehicles use around one third of the energy of their fossil counterparts. Electric heating also saves energy, as does electric cooking. The other invisible win is not needing all the waste energy generated when making our electricity with coal or gas. This nirvana is enabled by renewables.

As we've seen, there are four major energy uses in our homes today: vehicles, space heating, water heating and cooking. Let's dig into what we can expect to see in each of these areas as we electrify our homes. We'll see hip-pocket savings, zero emissions from energy, healthier living environments and a more comfortable future.

Electrify our garages!

What we have to win in electrifying our garages is lower driving costs. It will cost half as much, or maybe even only a tenth, to run electric vehicles (EVs) compared to our current belchers. EVs are arriving in force, with more than 20 new models of electric car expected in Australia in 2022. Prices are dropping fast, too, with Bloomberg NEF predicting EVs will be at price parity by 2026. This means an electric SUV, for example, will cost the same as a petrol SUV.[29] By 2030, EV prices are expected to drop to around 80% of the price of their petrol counterparts. It won't be long before electric vehicles are the obvious economic choice by far. Not only are electric cars cheaper to drive,

7.9: Comparison of driving costs by car type and energy source. Electric cars cost about half as much to drive when charged using the existing electricity grid, and cost about ten times less to drive when charged with rooftop solar.

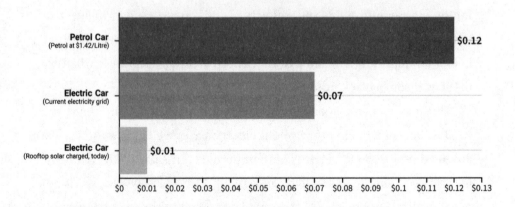

Source: Rewiring Australia, based on a petrol price of $1.42 per litre

they're also cheaper to maintain, requiring far less servicing than a hot, vibrating internal combustion engine containing hundreds or thousands of explosions per minute.

But let's not skip the fun parts. Yes, this is a win for your wallet, yes, this is a win for energy savings and the climate – but it's also a win for comfort, and a win for fun, as anyone with a Tesla will tell you. On hot days, you can schedule your car to turn on the air-con ten minutes before you leave for work; on cold days you can do the reverse and have your car warm and toasty the moment you hop into it. In tight parking spaces, you can get out of the car before parking, and then tell it (with an app on your phone) to drive into the parking spot for you – and to back out of it when you are ready to leave. The new Rivian ute has powerpoints throughout, so you can plug in your laptop, tools or just about anything. You can go on a camping adventure and power anything you might need, while a built-in air compressor can pump up your paddleboard, bike tyres or raft.

Electric vehicles are approximately three and a half times more efficient at converting energy into motion, and therefore unlock significant energy savings. Even charging an electric vehicle with the existing electricity grid will approximately halve fuelling costs, and if we charge our electric vehicles with rooftop solar, we are looking at around a tenth of the cost, as seen in Figure 7.9.

Electrify the living room!

Inside our castles, there are more savings to be had. Heating our lounge rooms, bedrooms and so on is the second-largest use of energy in the average Australian home, making up 11% of total energy use (or 37% of appliance energy use if we exclude vehicles). Conventional heating with gas, wood and resistive electric bar heaters is incredibly inefficient compared to electric heat pumps. Remember, a heat pump is just a reverse-cycle air-conditioner, something you may already have on your wall. Electrification of our space

7.10: Comparison of average daily heating costs by type and energy source. Electric heat-pump heating is about half the cost of gas heating using the existing electricity grid, and about ten times less when using rooftop solar. Using a battery to back up the solar and run heating at night is in between, and still far cheaper than gas heating.

Space heating cost comparison for one day – average home

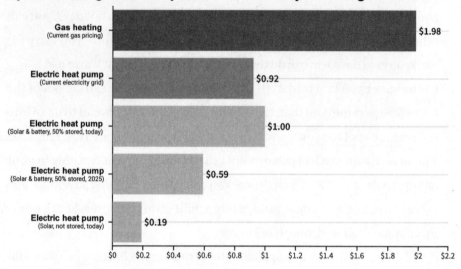

Source: Rewiring Australia

heating just means more of these modern air-cons, which can effectively regulate the temperature in our homes to whatever we find most comfortable. Most of them are automatic, and many can be scheduled to turn off when you're out and turn on just before you arrive home.

In Figure 7.10 I show the difference in cost between a gas heater, a heat pump using the current electricity grid and a heat pump using solar electricity. For the solar, I've also added a few extra bars to show what this looks like if we need to store that solar energy in a battery and use it at night. The cost to store 50% of that energy for use at night is just above electricity grid costs today – still much cheaper than gas, and with the forecast drop in battery prices by 2025 it will soon be even cheaper.

On the other side of the thermometer – the hot side that most Australians are very familiar with – we have space cooling (or air-conditioning and fans).

On average, space cooling in Australia requires significantly less energy than space heating. This means homes in the hottest parts of the country, like Queensland and the Northern Territory, use significantly less energy on average. Our current methods of space cooling are relatively efficient and are already nearly all electric. For these reasons we can leave them as they are in an electrified world, as long as we generate that electricity with renewables, and our air-cons will still get the cost benefits from cheaper solar electricity available in an electrified home.

Electrify the shower!

The future is not shorter showers. Instead – you guessed it – we are looking at cutting the cost of heating water at least in half, and possibly as low as a tenth of current costs. Water heating is the third-largest use of energy in the average Australian home, using 8% of total energy (or 24% if we exclude vehicles). Australia has a lot of gas water heaters – approximately 45% of current stock. Another 45% are electric-resistive, and the remaining 10%

7.11: Comparison of water heating costs by type and energy source. Electric heat-pump water heating is about half the cost of gas heating when the electricity comes from the existing grid, and about a tenth of the cost when using rooftop solar electricity.

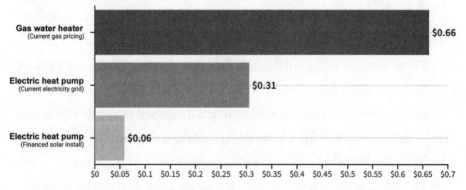

Water heating cost of one luxurious shower – Australian average

Source: Rewiring Australia, based on 3.93 kWh energy required, heat pump COP 3.78

are a combination of solar water heaters and heat-pump water heaters. On the existing electricity grid, the cost of water heating is roughly halved when using a heat pump compared to a natural-gas water heater. With financed solar panels providing the energy, the cost falls to around a tenth of that of gas heating, as shown in Figure 7.11.

The nice thing about water heaters is that they are batteries, and the water can be heated when the sun is shining or the wind is blowing. The use of water heaters (and space heaters) for energy storage and demand response is going to increase greatly and is one of the secret weapons we'll have in balancing a solar-heavy grid.

Electrify the spag-bol! (or stir-fry or falafel or stew)

Energy for cooking makes up approximately 2% of total energy use in the average Australian household (or 6% if we exclude vehicles). Electric cooking is generally at least twice as efficient as gas cooking, meaning more energy savings, more cost savings and more emissions savings. Perhaps the biggest hurdle is the psychological one, with many people having the impression that gas is the superior cooking technology. For now we'll skip right past gas companies paying social influencers to cook with gas and instead focus on the technological positives.[30] Modern induction cooktops are nothing like the old electric cooktops of the past – they're easier to clean and better at controlling and distributing heat because the heat is being sent directly into the pot (not just in its general direction).

Far more concerning are the impacts on respiratory health: gas cooking has been shown to make indoor air up to two to five times more polluted.[31] A recent report from the Climate Council discussed multiple studies on respiratory health risk from natural-gas cooking and likened the risk to children's health to having second-hand cigarette smoke in the home.[32]

The gas industry wants you to love your gas stove so it can make money selling you gas. They have sold us on the idea of the clean blue flame and the superiority of gas for cooking. Neither claim is true, and as more people experience modern induction cooking, its speed and easier cleaning, we'll be able to fight back in this insidious culture war. Electrifying our cooking is not only great for our wallets and the climate, it also benefits the health of our families.

7.12: Comparison of average daily stovetop cooking costs. Cooking with an electric induction stove is about 25% cheaper using the existing electricity grid, and about six times cheaper when using rooftop solar electricity. Storing that solar energy in a battery to then use at night is in between.

Daily stovetop cooking cost comparison – average home

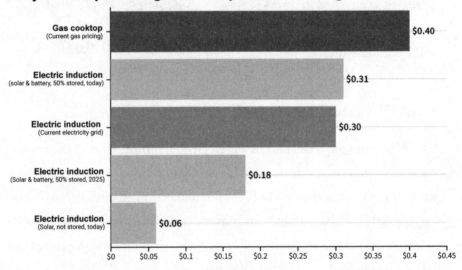

Source: Rewiring Australia, based on 0.80 kWh of base cooking energy required

Electrify the roof!

Australia leads the world in residential rooftop solar, with over 3 million homes basking in the free energy that lands on their rooftops each day. Rooftop solar already makes economic sense, with the cost of solar over its lifetime being far lower than that of grid electricity. Electrified homes will likely install even more and larger solar systems, as we electrify our vehicles and heating and cooking. Utility grids are already pushing back, claiming that the grid can't handle it. The grid can handle it – we can deploy more batteries and demand response and make it manageable with good engineering. No, it isn't what they've been lazily doing for the past five decades, but that's fine: even utilities and distribution networks need to adapt to the future. Australians own a stake in their distribution grids (which are typically at least partly owned by state governments) and we deserve them to work for us, not against us, including a neutrality policy that guarantees that your solar-powered mum in Mudgee is treated as an equal in terms of generation and storage to a power station in the Hunter Valley.

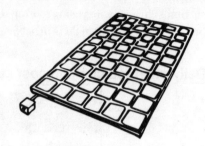

Electrify the side of the house!

Home batteries are a key part of our electrified future. Making the most of Australia's abundant sunshine will mean generating as much electricity as we can from solar in the middle of the day and storing that energy to be used when it suits us. Today, home batteries are very close to the crossover point of being an economically obvious decision when combined with a large solar install on households, and prices are dropping fast.

If you bought a residential battery in Australia in 2021, it would have cost about $1000 per kWh of storage (possibly more). I mentioned this in Chapter 4 but I'll add a reminder here for emphasis. Ford's new electric F150 ute, announced for 2022, is predicted to have a battery above 100 kWh and a starting price of under US$40,000. That's a battery at about half the price of current batteries, *and it comes with a free truck!* Battery prices are coming down fast, and this will help propel a boom in household electrification. With batteries at half the price of what they are today, the finance repayments for buying a large solar and battery install will be far less than average energy bills. You'll also be able to keep your house running as normal in a blackout or a disaster. An electrified future means both cheaper electricity and extra back-up security for your home.

In 2022 electric vehicles still seem like a luxury item, and participants in the culture war will try to use them as a wedge – but as these batteries in our cars and on our homes get cheaper and cheaper, the side of zero emissions will have the last laugh.

Electrify our other stuff!

Of course there are lots of other things in our home that consume energy. Luckily for us, most of them are already electric, which means they will benefit from lower running costs in an electrified home with rooftop solar. That cheap rooftop solar electricity can be used to charge your laptop and phone, and lower the running costs of your fridge, lighting, washing machine and TV.

There are a few other appliances that also use fossil fuels, such as barbecues, lawn mowers, leaf blowers and natural-gas pool heaters. While the climate change contributions of these may be small, they still stand to benefit from electrification and will likely soon be replaced with electric versions. Never needing to replace a barbecue gas bottle and never needing to fill up a jerrycan to put petrol in a lawn mower will be small wins for the climate, but perhaps bigger wins for the people who do these household chores.

Savings in the suburbs

Let's take a look at what electrification means for the average household budget, and what it means for the whole of Australia.

7.13: Spending and savings with home electrification and forecast technology prices. Electrifying an average home in 2022 is predicted to cost about $5500 extra per year (mainly from the capital cost finance from buying electric cars, home battery and solar). By 2025, electrifying the average home is predicted to save money, with those savings growing as technology prices get cheaper.

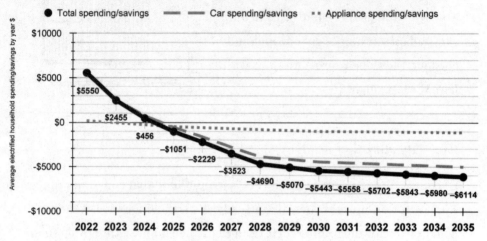

Aus. electrified household financed upgrade – spending/savings by year
Forecast savings for electrifying average household including financed cost difference of solar, battery, appliances, and electric

Imagine you need to replace all the devices in your home in 2022. These items will probably be due for replacement at different times, and therefore this cost will be spread over many years, but for simplicity let's imagine we are going to replace them all in 2022 – your space heaters, your water heater, your stove and your car(s). You could replace an item with something similar (a gas heater with another gas heater), or you could replace it with an electrified zero-emissions alternative. Figure 7.13 shows that in 2022, it will cost an extra $5550 for the average household to buy electrified appliances instead of conventional ones, including financing new electric cars, solar on the roof and a home back-up battery. This is the difference in replacement cost. It's worth noting that ongoing energy costs drop overnight, but the extra cost is the financed capital cost of buying all these new electrified devices. By 2024 we are nearly at break-even, with the financed cost of the hardware only increasing total energy costs by about $456 a year – a relative bargain for saving the Earth. In 2025 the numbers flip in favour of electrification: electrifying homes becomes the cheaper option and creates yearly savings for the average household of more than a thousand dollars per year. After that, helping to act on climate change becomes the best way to save money. And things only get better, with savings rising, meaning savings of over $5000 a year for the average Australian household in 2030.

You might ask, why start now, when the bigger savings will come in a couple of years? Which is a great question. The simple answer is that to get those savings later, we need to have first movers who drive the process forward. If no one is buying these things today, the price will never come down. To see those cost savings, we need to be manufacturing more solar panels and batteries than we are today, so that we figure out how to make them cheaper and scale up their manufacture. We need to be installing these

7.14: Cumulative investment and savings if Australia subsidised accelerated electrification of homes. A relatively small $12 billion investment to kickstart accelerated electrification could result in over $70 billion in savings by 2030, and over $300 billion in savings by 2035, all while significantly reducing Australia's emissions.

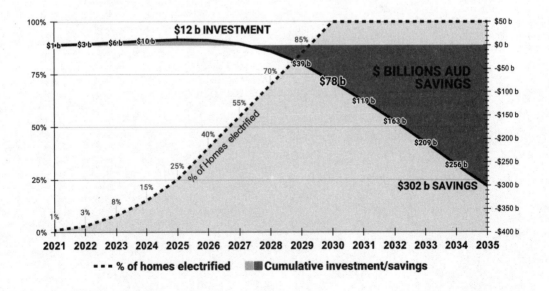

electrified homes so we can figure out how to do them at scale across the whole country. We need to do the work now so that by 2025 the average Australian family can electrify its home with ease. This is a traditional role for government – seeing the future, and through incentives and pilot programs buying down the cost of that future for its people as an investment in their future.

Figure 7.14 shows electrification spending and savings for the whole Australian economy. If we start electrifying homes today, we can figure out how to do so efficiently and effectively. We start with a small percentage of homes each year, gradually ramping up to all households in Australia. We electrify roughly 3% in 2022, 15% by 2024 and 40% by 2026, with a full 100% by 2030. Doing this would require an investment of about $12 billion and result in about $302 billion in household savings by 2035. It's a ridiculously good investment, considering it would also slash our emissions. For context, that's just roughly 9% of what we are predicted to have spent on the COVID-19 response in the last two financial years, or roughly 18% of our defence spending in those two years. Home electrification is an obvious win for Australia.

Chapter 8
Crushed rocks –
the export economy

- Australia exports a proverbial shit-ton of fossil fuels every year, for less profit than you may think.

- The future looks much brighter for our export industry if we start using the abundant renewable energy in our backyard to add value to our exported products.

- Australia is rich in the very materials the world needs for this energy transition, and we should invest heavily in that capacity, for our sake and the planet's.

Australia is an export-heavy nation, and the majority of those exports are rocks or other materials we dig up from the country's vast landscape. We export far more than we import, which leaves us with a surplus that is somewhat responsible for our high quality of life. Our exports also currently account for a very significant proportion of our emissions (as discussed in Chapter 2). Whether or not we 'count' these emissions as our own, they are nonetheless rocks that are dug up from our land and burned into carbon dioxide that affects ecosystems worldwide, including ours.

Our fossil-fuel exports are used by some to argue against climate action because of the money they make us. In this chapter I will argue that these export emissions are not only unnecessary, but that exporting different crushed rocks will be far more profitable for Australia's economy. I'm also going to build a basic argument for why our domestic abundance of zero-carbon renewable electricity can be used to add value to those crushed rocks by smelting them into metals and alloys and maybe even further processing them into finished products. Once upon a time we made cars. Remember that the rest of the world doesn't have our abundance of renewable resources and low population density and will be hungry to import low-carbon metals, fertilisers, agricultural products and perhaps even electricity and synthetic fuels (including hydrogen) from Australia.

Australia's dirty little fossil export secret

Today, Australia's exports are largely things we dig up out of the ground: 75% of our exports are minerals and metals, and most of the 'minerals' are fossil fuels. In 2019, 32% of exports were fossil fuels. The next biggest export category is agricultural, with large showings from meat, wool, wine and wheat. A detailed breakdown of our exports is shown in Figure 8.2. Our imports are another story entirely and can be seen in Figure 8.3. Our two biggest imports are petroleum (both refined and unrefined) and cars.

There is a persistent idea that our fossil exports are unequivocally good for our country. They create jobs. They generate income. But let's look at our dirty little export secret before we look at what we have to win if we think bigger about exports.

Fossil fuels are one of Australia's biggest commodities, but this is only one side of the ledger – we also import most of the petrol and diesel needed to fuel our cars. We export coal and gas, and we import oil products. Things become interesting when we start to balance out these fossil-fuel-related exports and imports. Australia exports roughly $130 billion worth of fossil fuels each year (these figures are from 2019), mainly coal and liquefied gas, and imports roughly $40 billion worth of petroleum and oil products. On the surface this may look like an okay picture (at least economically), but when we export a product, we have to pay all of the production costs for that product – mining it, transporting it, refining it and so on. So what we earn from our exports is the profit margin after we take out production costs; yet with imports we pay the full import price, which covers some-one else's production costs. If we estimate the profit margin of Australia's

8.1: Australia's dirty little fossil export secret. It's estimated that the profit margins on fossil-fuel exports barely or don't even cover the costs of importing the fossil fuels that power our vehicles.

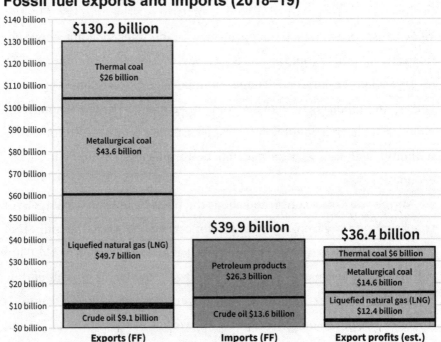

Fossil fuel exports and imports (2018–19)

Exports (FF) — $130.2 billion	Imports (FF) — $39.9 billion	Export profits (est.) — $36.4 billion
Thermal coal $26 billion		Thermal coal $6 billion
Metallurgical coal $43.6 billion		Metallurgical coal $14.6 billion
Liquefied natural gas (LNG) $49.7 billion	Petroleum products $26.3 billion	Liquefied natural gas (LNG) $12.4 billion
Crude oil $9.1 billion	Crude oil $13.6 billion	

Source: Department of Industry REQ June 2021 and Rewiring Australia

fossil-fuel exports, this amounts to roughly $36 billion. In other words, the money Australia earns from digging up and exporting all those fossil fuels each year barely or doesn't even cover the costs of importing the petrol and diesel to fuel our vehicles. If we remember that about 80% of the big mining conglomerates responsible for these exports are foreign-owned, then we actually only really get a fraction of those export profits and the picture is even more stark. In a net balance of fossil-fuel trade, the industry looks to lose the country money. Yes, it does create jobs, but the number is not that large. Yes, it might be profitable for the coal-exporting company, but all of us pay the price of not promoting renewables, electric vehicles and climate solutions.

If our vehicles were electric and were powered by Australian rooftop solar, we wouldn't need to import these fuels or export the fossil fuels to cover their cost. Protecting these fossil industries is often presented as a matter of economic conservatism, but that doesn't hold water: this is clearly not an economically good picture for the country. Electric vehicles are seen as sissy or inappropriate for our weekends, and hence un-Australian. Nothing could be more patriotic, though, than running your EV on Australian sunshine and making our imports–exports balance that much better by eliminating foreign oil from our economy. The scaremongering about switching our country away from fossil fuels is quite ridiculous when without fossil fuels we would be in a better place as a country in a basic economic sense.

Maybe you aren't concerned about the trade balance in fossil fuels, or transitioning away from coal, but are concerned about whether our other export commodities will be harmed or fouled in a zero-carbon world. They will. The EU is leading the world in planning border tariffs on high-carbon goods. The US is leaning in the same direction, and Asia will follow. If Australia doesn't clean up its exports the world will shun them, or flat-out make them worthless.

Australian exports 2019 – $284 billion USD

Source: Observatory of Economic Complexity

Australian imports 2019 – $209 billion USD

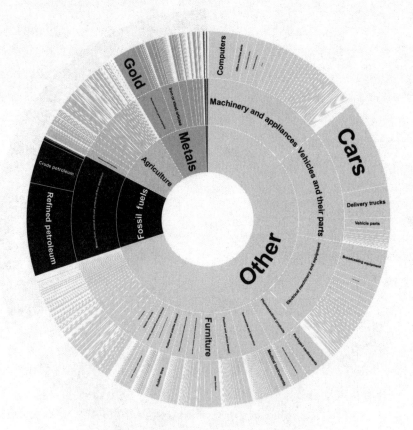

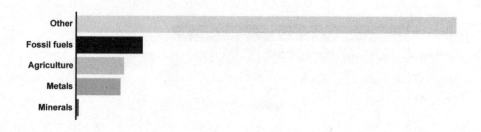

Source: Observatory of Economic Complexity

A superpower, yes, but which superpower?

I love that the concept of a 'renewable superpower' is now in the common parlance. I congratulate my friend Eytan Lenko and his buddies at Beyond Zero Emissions for introducing this idea in a report of the same name. 'Superpower' implies that we are much stronger than average, and hence can export energy. The question that deserves a little attention, though, is how that superpower manifests itself. The traditional method is to export something recognisable, like a fossil fuel, and it is this familiar tale that has many people thinking about how to export hydrogen. But another way to become a superpower could be by exporting electricity directly. Finally, one must also consider the giant opportunity to export energy as embodied energy, inside export products.

Abundance

The Australian backyard, with its long and colourful geological history, contains virtually every rock type from every period in Earth's history. Australia contains rocks that formed from volcanic activity just a few thousand years ago, along with rocks dating back more than 3 billion years. Table 8.1 shows a breakdown of the resources in Australia; we lead the world in reserves of gold, iron ore, lead, nickel, uranium, rutile, zinc, zircon and tantalum. We rank second in the world for bauxite (aluminium), cobalt, copper, ilmenite, lithium, tungsten and vanadium. We rank third in the world for reserves of silver and niobium, and we aren't far behind on antimony, manganese, tin, diamond, graphite and rare earth elements. Australia is abundant in the resources the world needs for a decarbonised future.

Table 8.1: Resources in Australia's backyard and where they rank on the world stage.
Source: Geoscience Australia

COMMODITY	RANKING	SHARE
Gold	1	21%
Iron ore	1	30%
Lead	1	41%
Nickel	1	24%
Rutile (titanium)	1	65%
Uranium	1	31%
Zinc	1	27%
Zircon	1	72%
Tantalum	1	73%
Bauxite	2	18%
Cobalt	2	19%
Copper	2	11%
Ilmenite	2	24%
Lithium	2	29%
Tungsten	2	11%
Vanadium	2	25%
Silver	3	16%
Niobium	3	2%
Antimony	4	7%
Manganese ore	4	14%
Tin	4	11%
Magnesite	6	3%
Rare earths	6	4%
Graphite	7	3%
Molybdenum	7	1%
Phosphate	9	2%

Becoming a more refined exporter

Australia's top-earning export for the past decade has been iron ore, with 82.4% of it going to China, 6.8% to Japan, 6% to South Korea, 1.8% to Taiwan and the rest to other countries. We also export significant amounts of other ores, like copper ore, zinc ore, bauxite and more.

The end product of iron ore is nearly always steel, and turning ore into steel is very energy-intensive: energy costs account for between 20% and 40% of the cost of making steel.[33] This was obvious to me in my first industrial job in the rolling mill at Newcastle. Huge amounts of energy went into heating the steel and rolling it into final products such as rails and rebar. My best mate was working in the blast furnace at the same facility – watching the pour was always a highlight of our week.

Australia currently exports roughly 89% of the iron ore we dig up, with only a small fraction being refined into steel domestically. It's worth considering whether we should be exporting these ores in such an unrefined state, or whether we could be creating a more abundant future for Australia if we exported them in other forms. Metaphorically, we make a lot of our export money selling flour, which other countries use to make bread – they then sell that bread for far more than the price of flour. Australia could be cooking our metals before we sell them and charging a much higher price. While iron ore sells for roughly $100 a ton, steel sells for roughly $1000 a ton. These numbers can change significantly with global economics, but at the end of the day the end product will always command the higher price. If other countries can afford to buy our iron ore, ship it and then use energy that is more expensive than ours to make steel, it follows that we could be making that steel here with our cheap renewable energy and selling it instead.

In Figure 8.4 we look at the current Australian iron-ore industry and compare it with what we could be doing instead as a renewable energy superpower. Today we pull 917 million tonnes of ore from the ground and export 818 million tonnes of it for around $78 billion in revenue. We use roughly 10 million tonnes of iron ore to make around 5.5 MT of steel domestically, about 80% of which is used in construction. We export a relatively tiny 1.2 million tonnes of steel. As a thought experiment, consider what it would be like if we processed all the iron ore into basic steel alloys here. We could be producing 665 MT of steel for export. This would take 15,000 PJ of energy, which is the amount of energy we export in fossil fuels today. This would

8.4: Comparing current iron-ore exports to potential exports of steel in an electrified Australia. Further refining our iron ore into steel with our abundant renewable energy could create a huge boost in export earnings.

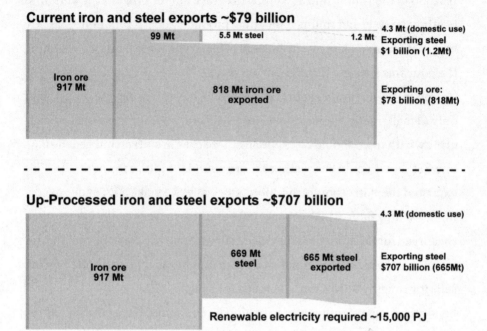

Current iron and steel exports ~$79 billion

| | | 99 Mt | 5.5 Mt steel | 1.2 Mt | 4.3 Mt (domestic use) |

Iron ore 917 Mt

99 Mt

5.5 Mt steel

1.2 Mt

4.3 Mt (domestic use)
Exporting steel
$1 billion (1.2Mt)

818 Mt iron ore exported

Exporting ore:
$78 billion (818Mt)

Up-Processed iron and steel exports ~$707 billion

4.3 Mt (domestic use)

Iron ore 917 Mt

669 Mt steel

665 Mt steel exported

Exporting steel
$707 billion (665Mt)

Renewable electricity required ~15,000 PJ

bring in $707 billion for the Australian economy, roughly ten times our current iron earnings, and nearly three times our total current export earnings.

Clearly it is not that simple, and it is obvious that the Chinese now have steel production over capacity and are dumping low-cost steel on the world's markets. On the other hand, there will be many who would like to buy 100% cleanly produced steel. And no doubt international carbon pricing will make this story better for Australia over time.

There are already two processes being developed to make steel without coal. Thyssenkrup is working with various partners on a process that replaces the coal with hydrogen. It is relatively low risk technologically and is already going beyond experimental scale. Boston Metals was started by a professor I used to know at MIT, Don Sadoway. They take an entirely

8.5: Comparing current bauxite, alumina and aluminium exports to potential exports of aluminium in an electrified Australia. Further refining our bauxite into aluminium with our abundant renewable energy could create a large boost in export earnings.

Current bauxite and aluminium exports ~$16 billion

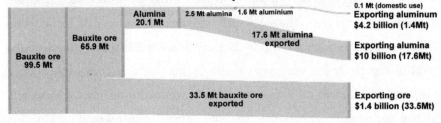

Up-processed bauxite and aluminium exports ~$48 billion

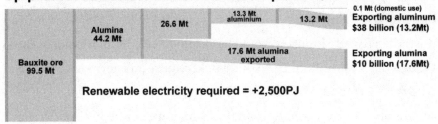

different approach, using electrochemistry instead of old-fashioned heat and beat. They are also already making ingots, if only at tiny scale. Bill Gates' Breakthrough Ventures and BHP have both invested. If I were a betting man, this is a really good-looking horse that is going to drop the price of making steel significantly and eliminate all the carbon dioxide.

Steel is not the only material we could export our renewable energy in. Let's look at another very energy-intensive metal, aluminium. Aluminium is made from alumina, which is extracted from bauxite. Australia has a lot of bauxite and our current exports of all things aluminium are shown in Figure 8.5 We currently mine close to 100 MT of bauxite and export 33.5 MT of it as ore, for a comparatively low $1.4 billion in revenue. The remaining 66 MT of ore is turned into alumina in six refineries in Western

Australia and Queensland. Most of that alumina (17.6 MT) is exported, for $10 billion in revenue. The remainder is smelted into 1.6 MT of aluminium in four smelters, one each in Victoria, Tasmania, New South Wales and Queensland. Most of this is exported, for $4.2 billion in revenue. Most of the emissions from making alumina are from this smelting process, where the carbon electrodes in the electrically driven process burn off into carbon dioxide. Apple, in partnership with Alcoa and Rio Tinto, has developed a carbon-free, electrified pathway to achieve this smelting. Should we choose to export all-electric, all-renewable, zero-carbon aluminium to the world, with all the bauxite we produce that isn't currently earmarked for alumina exports, we'd be turning our 99.5 MT of bauxite into 44.2 MT of alumina, selling 17.6 MT of that for $10 billion and converting the remaining 26.6 MT into 13.2 MT of aluminium, worth a whopping $38 billion. This version of an Australian bauxite industry would be earning $48 billion in exports, three times today's $16 billion. To do this would require an enormous 2500 PJ of renewable electricity – making us a superpower.

Metals aren't the only product in which we could be exporting our clean renewable energy. The world needs fertiliser, and lots of it. Most of the world's production of 60 million tonnes of hydrogen today goes into ammonia, much of which is used to make the fertilisers essential to food production. The hydrogen for this market is worth about $90 billion. Given how dirty hydrogen produced from natural gas is today, making the hydrogen for the world's fertiliser from renewable sources is critically important. Today it takes about 50 kWh to produce a kilogram of hydrogen. If we do it as well as thermodynamics will allow us, it could take as little as 39 kWh. But we aren't that good at making hydrogen yet, so 45 kWh per kilogram is a more reasonable estimate for the green-hydrogen industry. If Australia was to produce all the world's hydrogen for ammonia, it would require around 9500 PJ of energy. Processing that into 144 MT of ammonia using something like the Haber-Bosch reaction would take an additional 6500 PJ In the unlikely case we make all the world's ammonia, we'll need as much renewable energy as we currently export in all our fossil fuels. Clearly this is also an opportunity for a clean-energy superpower.

You can see where this is going. Australia has an enormous opportunity to export energy to the world in our mined, processed and, I hope, electrically refined metals and other physical commodities, including fertiliser. If we committed to doing even a portion of this, we could be exporting far more energy than we are today and generating far higher revenue. One of the unrecognised benefits of being a renewable-energy superpower is that when those giant clean industries are connected to a nationwide grid, their energy requirements will dwarf those of the domestic economy. Given those industrial processes have large thermal inertia (that is, it takes a long time to heat up steel and aluminium and even longer for them to cool down), these giant industries can in effect become a huge battery for the rest of the economy – subsidising low-cost, reliable electricity and making our pathway to zero that much easier.

Exporting sunshine

The company Sun Cable has been working hard to bring a giant new export industry into life. It wants to export Australian solar energy directly to our neighbours in Southeast Asia. Enormous arrays in the north and west of Australia, where the world's best solar resources lie, pump electrons through giant underwater cables to the grids in Indonesia and Singapore, where the population density is high enough to make renewables (and nuclear) a difficult proposition. Sun Cable intends to provide nearly 3 TWh of electricity per year, or 95 PJ, in its first deal with Singapore, and that's only the first project. It requires thousands of kilometres of transmission lines and underwater cables, in addition to novel solar installations provided by an Australian startup, 5B. It is heroic and audacious, but there is nothing enormously technically risky about the project. There are political risks and security risks, but there is no reason to believe this won't be a proud Australian export, and the first of many.

Exporting fuels

I haven't been hugely kind to hydrogen as a fuel, but for a host of reasons, including non-technical ones like national security, some of the world's energy needs will be met with hydrogen. Beyond hydrogen for fertiliser, if Australia wanted to export hydrogen produced via electrolysis from low-cost wind and solar, Western Australia would be a pretty fabulous place to do it. Even if it is only a tiny fraction of the world's energy supply, perhaps 1–2%, it would represent at least 5000 PJ and be a substantial portion of Australia's energy production. This is roughly equivalent to the energy used to power international shipping, international flights, or heavy mining vehicles and freight rail around the world. It would represent $100–200 billion in revenue. This could be a great business and Andrew 'Twiggy' Forrest has already put up his hand to become Australia's hydrogen titan. Does it mean the future world economy will be a hydrogen economy? No. Does it mean Australia could make a motza exporting hydrogen? Yes.

The world's foundry

Australia is the sixth-largest country on Earth, and has the third-lowest population density, exceeded only by Namibia and Mongolia. We also have an outback filled with sunshine, wind and flooding rains – or in other words, solar, wind and hydro energy. What's more, because of our low population density, we have relatively small energy needs compared to countries with higher populations.

If our domestic economy story is one of savings in the suburbs and putting money back in everyday Australian pockets, then our export opportunity is one of an enormous number of jobs in the regions, money in exports and a pathway for Australia to continue its incredible luck as a primary producer of abundant resources – only now in an Earth-healing, rather than an Earth-destroying, way. We have the steel and aluminium that will be needed to modernise and electrify global infrastructure. We have the lithium, cobalt and nickel that will be needed to produce vast quantities of batteries for electric vehicles and storage.

The world already needs a lot of metals. In 2019 we mined around 3.2 billion tonnes of metal ore globally.[34] For context, that's about 60,000 times the weight of the steel used in the Sydney Harbour Bridge. It's the equivalent of building roughly 300 brand-new Sydney Harbour Bridges in every single country in the world, every single year. The world is going to need a lot more metals as low-income countries build vast amounts of new infrastructure, and as middle- and high-income countries rebuild their infrastructure for an electrified, decarbonised world.

The clever country

When I was a teenager in Australia in the 1980s, we had great aspirations to be the clever country, to manufacture more domestically, to export our inventions to the world. It didn't pan out that way. I left Australia to go to America to attend the Massachusetts Institute of Technology, arguably the best technology university in the world. There I saw what real innovation, ambition and national investment in its best and brightest looked like. Australia doesn't punch as far above its weight as it thinks it does in innovation, but it could, and it should.

Right now, on the back of the rooftop solar revolution, Australia is facing challenges pertaining to heavy renewable energy integration that other countries are years away from facing. We are a nation of practical inventors who find ways to solve these types of problems. Being first in the world to get our suburbs to go all electric, to figure out how to integrate the vehicles and the electrified heating systems, and to pioneer the distribution grid-level inventions that will tie it all together more cheaply – these are huge opportunities for Australia to lead the world with the innovations that will make the 21st century tick. These could be our fast-growth startups and export companies of the future.

And here is a radical idea: we could make stuff again. We could make batteries, we could make offshore wind platforms and turbine towers. We could even make cars, or parts of cars. We could make heat pumps and world-beating induction stoves and energy management systems and new transmission-line innovations. The traditional excuses for Australia's underperformance in technology and innovation – we are too small, or too far away – no longer hold. The internet has levelled the playing field.

I'm excited for Australia: if we choose to lead the world, we'll build the world's leading technologies and technology companies. But my fear for Australia is that we'll choose instead to stick with fossil fuels and be the fools who let the industries of the future slip through our fingers.

Chapter 9
Why politicians and regulations matter

- The right government policy can help us save money, slash emissions and get to an abundant future faster.

- We lead the world on rooftop solar policy – in regulations, certification, and financing.

- We need similar leadership on electric vehicles, heat pumps, induction and electric cooking, and batteries.

One key message of this book is that Australians are leaders in at least one aspect of the global clean-energy revolution: saving money with solar. In a few sweet years we can get to zero emissions in our households and automobiles at no cost. Better than no cost, actually: with savings.

Good government policy is critical to this big switch. The modelling presented in this book demonstrates that Australia can build on our household solar miracle to lead the way to full household and vehicle electrification. We have the most to benefit and we can get there first. The argument of this chapter is that we should start by discovering how clever policy enabled the rooftop solar revolution in the 2000s and apply those lessons to the 2020s. We might like to think that governments don't matter, or that they are all the same, and for sure we don't like to think about rules and bureaucracy and regulations and red tape, but this is where the front line of the climate battle will unfold. If you want to do something about climate change, don't chain yourself to a fence; engage with your local council.

There will be many other areas of policy that we need to improve, and we'll also look at those. It is after all a climate emergency; we need to pull every lever that makes a difference, and our rules and regulations are some of the largest. Some levers are emergency brakes (mandates and moratoriums); some are incentives (subsidies, tax deductions); some are disincentives (taxes); and some are dials to be turned up (education and training) or systems to be optimised (building codes). It is critical that we look at every option we have to speed our transition to a better, electrified Australia. If the most important chapter to read was Chapter 5: Electrify (Almost) Everything!, for better or worse this is the second-most important. Without the right policy settings, everything else is either motivational storytelling or fancy graphs.

Given the scale of the task we're facing, every single policy that can enable our transition needs to be on the table. The federal government should lead with a national plan for electrification, channelling the leadership of state governments and coordinating a national effort. The government should fund pilot programs, take the lead on planning and building reform; organise training for industry; solve supply-chain bottlenecks; and provide rebates and concessional finance. The market is imperfect and can't get the job done on schedule, so it needs some help. If your ideology has a problem with that, I have a problem with your ideology. It would be nice if our federal government were a global leader on a rapid climate transition – it sure should be.

Governments can determine who shares in the incredible economic benefits that will flow as we replace expensive dirty fuels with cheap clean renewables. As in any revolution, there will be winners and losers. If we are smart, we'll try to ensure that everyday people are winners. If we are even smarter, we'll be thinking about how to help the potential losers (the owners of coal, oil and gas assets) become winners by transitioning their organisations to the future rather than clinging to the past. The current political narrative claims that workers who mine coal and tap gas are going to be losers, but without doubt there is a place in the green economy for people with their skills.

In the US, an organisation I started, Rewiring America, helped to form an Electrification Caucus – a growing group of congressional representatives putting electrification at the top of the American political agenda. At our launch event, Representative Sean Casten of Illinois described the upcoming energy transition away from fossil fuels as 'the greatest transfer of wealth in human history, from energy producers to energy consumers'. That is an extraordinary statement, but I agree. Given the household savings we have discussed, there is a lot of money on the table for someone to grab. Government policy will determine who the winners are. We can already describe broadly the four groups that stand to benefit: (1) households, (2) utilities, distributors and retailers, (3) banks, and (4) third-party providers who sell managed energy services, leases and other pathways to fossil freedom. There will need to be a little money in it for everyone for the system to work, but I think there should be a clear government strategy to write the rules to favour households, the fundamental economic unit of the economy.

Cutting through the green tape

Three million Australians households have a little solar power station on the roof of their house. At 30% penetration and rising quickly, it now seems obvious that we would be the world's solar leader, given we have so much sun, an urbanised population, and big houses with plenty of roof space. But it took more than that, and Australia's solar uptake is a case study in regulatory success. Without good policy that unleashed solar competition, we would not be where we are today. There's plenty of credit to go around. Some is due to a hero too few people know, some to a scientist well known in the energy world but not outside it, and part of the amazing untold story is the role played by a conservative prime minister, John Howard.

In the early days of solar in the 1980s and 1990s, it was a technology for visionaries, not suburbanites. The technology was complicated and the only people who got PV on their houses were renewable energy nerds – the types who planted native gardens, recycled and composted, rode to work on their Malvern Star ten-speeds and saved up for expensive photovoltaic systems that nobody else really understood.

During this period, the solar scientists at the University of New South Wales, led by Dr Martin Green, made many of the innovations in silicon PV that have turned it into the dominant technology that it is today. Some 90% of the world's solar modules are based on technology developed at UNSW, out of his laboratory in Kensington. The UNSW scientists decided to install a solar system at the university, so they started by asking the electricity network how to go about getting permission to use their own technology on their own roof-space. Solar was so new at this point that the application form was the one used for large coal or gas generators! The solar system the scientists were proposing to install was about one 150,000th the size of a coal power station.

This was a perfect example of what might be called 'green tape': old rules designed for a coal-centred world that hold back renewables and push up the cost of energy. Fossil fuels have dominated our energy supply for 150 years, so they've had 150 years to write rules and regulations in their favour. I don't think it is a conspiracy – it was undoubtedly a good idea at the time, but that was before we realised that these fossil fuels were killing us and our coral reefs. By the 2000s, solar was starting to become a cost-competitive way for ordinary households to save money. The upfront cost was still expensive, as it was being manufactured in high-cost places such as Germany, the US and South Korea. Industry pundits knew the price would fall, and that government incentives would help the industry to get to the scale that would make it a money-maker, but no one knew just how much China's embrace of solar manufacturing at scale would change the game forever.

In America, the supposed bastion of efficient capitalism, solar households are burdened with bureaucracy, which pushes up their costs. In America, if you want solar, first you need to get a permit to put it on your roof, something that can take weeks or months. Then the solar company visits your house twice, once to check the roof, next to install the system. Then the local government's representative has to visit to approve the system's compliance with state and federal safety standards. Next you have to strike an agreement with a probably obstinate energy company to hook your solar into their network – and only then can you start generating energy.

Thankfully, Australia learned the lessons of those early UNSW experiments and cut through green tape. Engineer Geoff Stapleton and some colleagues were obsessed with making solar cheap and safe to install, and so they created Global Sustainable Energy Solutions (Australia), or GSES. Their contribution was to train people and build the vital human capacity required to install a huge number of small solar powerplants. In Australia, it can take just one site visit for the installation of your solar system. You can order solar online: the installer uses satellite images to check out your roof and design a system remotely. The same electrician who installs the system does the required safety checks and approvals. That last bit is important – by standardising training, Geoff, GSES and the federal government eliminated many of the 'soft costs' that plague rooftop solar in the rest of the world. For perspective, and using round figures for 2020, the solar modules themselves cost 25 cents per watt. In Australia, the installed cost is one dollar per watt, while in the US ('land of the free'), it is three dollars per watt.

It was the Liberal federal government of John Howard that designed this elegant regulatory solution. As the costs of electric vehicles and household electrification come down over the next few years, the question will be whether the existing rules hold back or encourage households and businesses to clean up their act and save money with solar. Political parties should be competing to design policies that put the most money in the pockets of regular punters. Those punters vote, after all. The good news is that there is a political consensus in Australia around removing barriers to clean energy. Liberal, Labor and Green politicians have all backed reforms that would open up the electricity market. But we need them to move faster, and be bolder. It's not like climate change can wait.

The time is right

Since forever, climate and other ecological issues have seemed like a terrible trade-off. You can save money or do good – you must choose! Politicians, media and experts keep hammering the dilemma: jobs versus the environment. But we have already arrived, or are about to arrive, at a time where you don't have to choose. I am an inventor and I have never doubted that humanity can invent clean ways to cook a chook or chill a chardonnay, to keep Nanna warm and drive the kids to school. Clean energy is a win-win-win: it can be plentiful and cheap. Installing all the new gear will create jobs and economic activity. We can have our cake and eat it too.

So does this mean that we can stand back and Australia's electrification will miraculously happen? No. Vigilance is required. It might happen eventually, but we need the transformation to occur at unprecedented speed.

Households and investors will end up financing most of the cost of rewiring Australia, but governments have to help. They can do this in three significant ways: (1) stopping subsidising the problem (i.e. fossil fuels); (2) eliminating green tape to make the task cheaper; and (3) focusing government money on speeding this energy transition through pilots, incentives, rebates and subsidies that help the nascent market develop.

Governments wrote the rules of the grid, and those old rules were designed around coal. Government built the institutions that govern the grid. These institutions are staffed by engineers with a conservative mandate not to rock the boat. We need more risk-tolerance than that.

We have succeeded in lowering the price of solar, but don't be fooled: Australia's electricity laws are still a hidden subsidy to fossil fuels and a barrier to clean energy. Old, coal-centred policies are a tax on households that want to do the right thing. The risk-averse distribution companies that get the electricity from where it is generated to your home are worried about their mandates to make electricity reliable, and are slowing the transition to more distributed energy and storage (the solar on your roof and batteries in your garage) because that is their business. Australians are lucky, though: the people still own about a third of the networks, generators and retail. Queenslanders own their entire grid. We shouldn't sell it to the private sector; we should use our ownership to push energy companies to do the right thing, the thing we want and need, which is to electrify and put 21st-century distributed generation and storage on an equal footing with their 20th-century centralised predecessors.

Governments around the world spend trillions of dollars on subsidies that prop up dirty energy. But clean energy cannot sweep away dirty energy if it can't compete in the market fairly. We got the policies and capacity building right in Australian rooftop solar, but we are still handicapping electric vehicles, home batteries and even larger rooftop solar systems.

The Australian government has a key role to play in encouraging its energy innovation agencies to fund pilot electrification projects in cities and regions. But those agencies are busy and distracted with a lot of things, including carbon sequestration, agriculture, industry, hydrogen and other technologies that, compared to electrification, seem to me to be merely headlines or 'announceables', not substantive solutions. We can't do the whole energy transition on our rooftops. For reasons of economics, security and reliability, we will always need a mix of distributed (rooftop-scale, off the grid) and centralised (industrial-scale, on the grid) infrastructure. Let's not pit the two alternatives against each other, but rather work to find the optimal balance. State governments will have a vital role here, in building renewable energy zones in regional Australia that will host hundreds of large solar, wind and battery projects. These will replace the energy (and jobs) provided by our ageing coal fleet and allow it to retire quickly.

In the past, local governments have owned electricity generation and supply. That might not be a bad idea to revisit. Certainly, having more experiments running, testing all the details of suburban electrification, would be to everyone's advantage. A small number of trials run by national agencies can't compete with a large number of councils figuring out what works best. Local councils are already making a huge contribution to building a low-emissions future for Australians. They can help citizens and ratepayers work together with businesses and civil society to develop a shared zero-emissions plan for their village or town or suburb. The idea that they could run their own grids again should at least be considered. Things change. No policy should be irreversible.

If everyone who reads this book becomes an advocate for electrification, it will happen faster. Your local council and the members of state and federal parliament who represent you need to know what future you want. Do you want billions of dollars subsidising dirty energy in some far-off electorate because it wins the incumbent party a few critical votes, and crazy grid rules that make it hard for people to go solar? Or do you want cheaper energy and a safe climate for your children?

Fossil-fuel subsidies and federal policy failure

National climate policy in Australia has been brutal for more than a decade. Australia's net-zero target for 2050 lacks any detail to show how we will reach it. It implicitly relies on things such as negative emissions and hydrogen that are either easy to assess as physically unlikely or just plain fantasy. Our 2030 target is inadequate.

One policy in particular deserves to be called out: fossil-fuel subsidies. One of the worst indicators of our national policy vacuum is the fact that our federal government spends more money making the climate emergency worse than it does trying to fix the problem. We hide this behind a 'technology, not taxes' mantra of free-market optimism, but subsidising fossil fuels is the antithesis of a free market, and the reason governments are elected is to make wise decisions that are beyond the capacity of a market – like saving our reefs and protecting our children.

One recent study calculates that around $10 billion were spent by governments in Australia to prop up fossil fuels during financial year 2020–2021. This was mostly from the federal government. The biggest item was the $7.8 billion fuel-tax credit. This went to major users of fossil fuels, including $1.5 billion to coal and gas producers. That means taxpayers are paying fossil-fuel companies to rip more carbon out of the ground and accelerate the climate emergency. There was another billion dollars spent on aviation fuel and offshore petroleum tax concessions, and $ 1.4 billion in support for fossil-fuel industries, much of it from state governments, particularly Queensland's.[35] Unfortunately, we are not unique. According to the OECD estimates, fossil-fuel industries globally benefited from $6.9 trillion in subsidies between 2010 and 2020.[36]

States for the win

When state power and independence works, it works well. Australia's federation has the potential to be an incubator for the best electrification policy in the world. Solar is plentiful and grid electricity and gas are expensive, which makes electrification economically attractive. And because each state is its own electricity fiefdom, we could have a race to the top in policy. This comes about because the Australian constitution left the states in charge of electricity. Strangely, the federal government has no direct authority to decide how the electricity grid works. The states are responsible for granting the right to build electricity networks and generators, and states license the electricians who install the batteries.

The federal government is supposed to set the national agenda on climate policy, but it has done such a bad job that the states have stepped into the breach. Our state and territory jurisdictions have set net-zero targets and plan to deliver them. This effectively means that Australia does in fact have a plan, because the country is made up of all those states and territories. The collective plan of the states is in conflict with the position the federal government represents to the world. These jurisdictions are already competing with one another to best manage the clean-energy revolution and they are doing a pretty great job of it. Over the next few years, they can learn from each other and accelerate decarbonisation, starting with households, the grid and cars.

Different states have different challenges and have innovated their own solutions. This section is a summary of the best of them – but it is a dynamic space, and by the time this is published I'm sure more good state policies will have been announced.

New South Wales

The richest and most populous state in the country has a reasonably ambitious climate plan. The goal is net zero by 2050. As mentioned earlier, the 'net' in net zero is a bit of a cheat, and because we are a wealthy nation and could go faster, this target should really be more like absolute zero by 2035 or 2040, which better accords with the science of hitting 1.5°C. NSW is taken seriously because its government has released a detailed plan for the first decade of the strategy. It intends the state to achieve half of its emissions reduction by 2030. Most of the reduction over the 2020s will come from renewable energy displacing coal for electricity, and industrial innovation around efficiency and electrification. There is $195 million for technology innovation, including pilot schemes.

The most visible and dramatic part of the strategy is to build enough solar, wind and storage to replace coal power stations, which currently generate around two-fifths of NSW's electricity. This should be applauded but, to be clear, the government has not modelled how rapid electrification will approximately double the amount of electricity we need. Very few governments recognise yet that electrification – our only decarbonisation strategy – requires not just replacing retiring coal plants, but massively increasing supply.

The federal government is supporting NSW to host the national pilot Renewable Energy Zone in the middle of the state. By the mid 2020s, the Central West–Orana Renewable Energy Zone will host about 3 gigawatts of mostly solar and some wind generation, enough to power 1.4 million homes. The zone was chosen because it has lots of sun and plenty of strong wind and sits near existing high-voltage power lines that can carry power to the east coast cities and industrial centres. By the 2030s, there will be Renewable Energy Zones in many other states as part of the National Electricity Market.*

New residential properties will have to meet a higher thermal energy rating of seven stars, up from 5.5 currently. One of the headline impacts of this new policy is that dark roofs are out in NSW. Dark roofs absorb solar energy and radiate it in the local community, causing urban heat islands. Even better would be to mandate rooftop solar on all new builds and on significant remodels or retrofits. As a physicist, I like the simplicity of light roofs as a climate adaptation policy. A lighter-colour roof makes it cheaper to keep a building cool in summer and reduces the heat island effect, reducing the ambient temperature for everyone. As global warming increases heatwaves, future generations of urbanites will be grateful for lighter roofs and cooler neighbourhoods.

South Australia

South Australia made negative headlines in 2016 when it had the first blackout since the National Electricity Market was formed two decades earlier. After this shock, SA put in place systems to make its grid more resilient,

---

*   The National Electricity Market is misnamed, as it excludes WA and NT. When
       I refer to reforms and policy measures around electricity markets in this book,
       I am generally referring to the NEM states and territories.

which enable clean energy to replace fossil fuels. In 2017 the world's biggest battery was switched on in SA and it has helped keep the grid stable. The 100 MW, 129 MWh Hornsdale Power Reserve is justly famous as a piece of technology, but the state government should be famous for its innovative policy support that made the battery possible. The government buys grid reliability from the battery. This 'frequency control' is a public good and the state buys it on behalf of all consumers, which is very efficient. (Frequency control keeps the alternating current in the grid very close to the target 50 Hz that electronic devices like your laptop and big machines like factory engines and agricultural pumps need to function properly.) SA also installed 'synchronous condensers' in its grid, to maintain the voltage within safe limits. These 'syncons' are essentially the electricity-generating part of a power station, but without the fossil-fuel power generator to drive it. They spin freely and match the 50 Hz grid frequency and maintain both frequency and voltage control.

The result of these and other innovations is that SA keeps breaking new clean-energy records. Last year it became the first gigawatt-scale grid in the world to almost entirely decarbonise. The syncons and big battery allow it to reduce how many gas generators are paid to maintain security. For brief periods of time, all demand in SA is met by rooftop and large-scale clean energy. This is amazing progress for Australia and the world.

Victoria

Victoria has created a dedicated agency to electrify households. Solar Victoria has a $1.7 billion budget over a decade. It helps households, particularly low-income ones, buy solar and battery systems and replace outdated space and water heating with efficient heat pumps.

Victoria is also leading the country with offshore wind. Wind is stronger and most constant over oceans. There is also less conflict with other ecological values or other human activities; there are not a lot of people at sea!

Turbines at sea are giants. They can be as high as 260 metres and produce as much as 14 MW, enough to power 3500 homes. A lovely quirk of history is that the first offshore wind farms proposed for Australia will be located in Bass Strait. They will generate energy in a sea that was used for offshore gas, and the power lines will come onshore and plug into the grid in the La Trobe Valley, the brown-coal powerhouse of the nation. This switch from coal and gas to clean energy perfectly symbolises how Australia can lead the world this century.

Queensland

Queensland has become a leader in using household energy devices to keep the grid running smoothly. Half of the consumers in the state have given the grid operator a way to remotely control one or more devices in their home. If the electricity system is going to run out of power for a short period and supply cannot keep up, then Queensland can briefly turn down demand. It only happens a few times each year and is fully automated. This is demand response, and according to the International Energy Agency, it will be a critical tool to help economies hit the 1.5°C Paris Agreement target.

Some pro-coal commentators have attacked demand response as if it means people have to give up consumer comforts, but that is not true. Nearly 100,000 households in Queensland have their heat pumps wired up to the Peak Smart demand response system. If the grid is under strain, the operator can turn down the compressors for up to half an hour but keep the fans blowing. A survey in 2018 revealed that over 80 per cent of Peak Smart consumers who had their air-conditioning throttled during the heatwave that summer would recommend participating in the program. Engineers describe demand response as a 'virtual power plant'. Instead of building a big new power station to be available for a few hours of peak demand at the height of summer, we can wire together millions of small consumer devices to use slightly less energy during a heatwave.

Queensland has also hosted innovative projects that show how solar, wind and batteries can maintain grid security better than conventional machines like synchronous condensers. In northern Queensland the inverters that connect four large solar farms to the grid were reprogrammed and now keep that part of the grid secure, at only 4 per cent of the cost of a conventional solution.

Tasmania

The island state of Tasmania is close to 100 per cent renewable energy because of its hydroelectric power and has a plan to be 200 per cent renewable by 2040. The idea is to use a combination of hydro and wind to generate twice as much energy as Tasmania needs on an annual basis and export energy to the mainland through undersea interconnectors. Tasmania can be the world pioneer in connecting grids across open waterways with cables, making everyone's grid more robust with its hydro, which could be used as a pumped hydro battery.

Northern Territory

The Northern Territory has around 400 remote Aboriginal and other communities that are a long way from the shared electricity grid.[37] These 'microgrids' rely on small generators powered by diesel, which can cost up to two dollars per litre in the outback. John Howard's government funded the Bushlight program in 2002, which installed solar and battery systems in 150 remote Indigenous communities. Last year the First Nations Clean Energy Network was launched and it builds on these successes. Aboriginal communities and entrepreneurs have developed ambitious plans for Indigenous Australians to take charge of their clean energy future. This is an incredibly inspiring movement, with some of the most remote and disadvantaged communities managing their own energy with large clean-energy cooperatives owned by First Nations communities.

Western Australia

In Western Australia the state-owned utility has gone one step further. As well as adding solar and batteries to microgrids to reduce diesel use, the government has worked out that it is cheaper to cut the lines to some remote towns at the edge of the grid and create microgrids, based on solar and batteries. These disconnected microgrids of up to 600 kW peak capacity are part of a new modular strategy for the WA electricity system.[38]

The national agenda

With all this innovation going on at state and territory level, it is clear that we can double down on clean energy this decade. The rest of this chapter will set out the key things Australia can do to turbo-charge the clean-energy revolution and electrify our castles this decade.

First, I show how governments can use pilot electrification schemes in suburbs and towns to develop the technologies and business models and create an export market. I then explain how Australia can cut the green tape holding back electrification and shift subsidies from the bad to the good. The next section looks at how we can use policy to encourage green buildings, which are ready for the electrified future and designed to make the most of the ambient energy flows, to increase comfort and reduce costs.

Australia's love affair with rooftop solar will show how the old coal-centred rules of the market create barriers that hold back household solar energy. Solar energy is the perfect people power and I next give you a summary of all the policies you can get behind, to expand your right to sell energy and services over the grid and invest in cooperative energy and community batteries.

Pilots to guide us

Full household and vehicle electrification will be cost-effective from around 2025. If we want to ride the wave and lead the world, we have to get ready. Governments must fund electrification pilots around Australia, to work out how best to put all the pieces together.

A major technology transformation like rewiring a nation is not just about the machines. It is a fundamental shift in complex systems involving economics, government, trade, technology, business models, politics and culture. There is only one way to figure out how it all works – give it a go! The years 2022 to 2024 are the window of opportunity for Australia to become the electrification powerhouse of the world. The centrepiece technology in this revolution is PV, and Australia is the world leader both in research and adoption. The economics of electrification are better here than anywhere else.

Pilots are a vital part of innovation. Governments can play the lead role by subsidising pilots that demonstrate to the public, investors and manufacturers how the technology works in practice, and what the benefits are. There are some individual homes that have gone all-electric in Australia. Electric kitchen, electric heat, solar roof, battery, electric vehicles and energy management systems. These early adopters have paid for the privilege, because to date this has been a pain in the arse. Just getting the contractors and tradies to do the right thing is tough when you are asking them to do something for the first time. Multiply that challenge by the entire supply chain and regulatory environment. While these brave individuals have made it work on a single home, there isn't a single street where everyone has done it, and definitely not a single suburb that is electrified and decarbonised. This is the front line of climate ambition, and this detailed integration is where Australia has the opportunity to shine.

One location, all the things

A full electrification pilot means rewiring all the households in one area with all the electric things. Each freestanding house would have solar on the roof. One car in each household would be a plug-in electric vehicle. Everyone would have a battery. Efficient heat pumps would replace all the old gas or electric hot water and space heaters. Induction stoves in every kitchen. Plus clever energy-management systems at the switchboard and an easy-to-use control app on everyone's mobile.

In practice, some households will refuse to participate, especially in the first trials. This does not invalidate the value of the research. As long as more than 60–70 per cent of freestanding households enrol, the pilot will be a rigorous test of electrification. The minimum grouping of households would be the 50 or so houses that take power from the same distribution transformer. You might have noticed these devices mounted on electricity poles. They are about the size of a bar fridge and have prominent wires coming in and out, with big insulation disks around them. A larger pilot would aggregate households connecting to adjacent distribution transformers. The goal would be to scale a pilot so that it included every house connected to the neighbourhood substation. Substations each feed power to about 1000 homes. The components can be seen in Figure 9.1, which shows how the house is connected to the 'street level' and 'transformer level' of the distribution grid. A large-scale electrification pilot would encompass the 'suburb level' up through the 'distribution level'.

By connecting all the homes in a street or suburb, these pilots will do much to prove that climate action makes environmental and economic sense. They will help create the cultural momentum for the switch to clean energy, as people share their experiences with their neighbours.

9.1: Pilot scales. The gold standard for a large-scale electrification pilot would encompass the 'suburb level' or 'distribution level'.

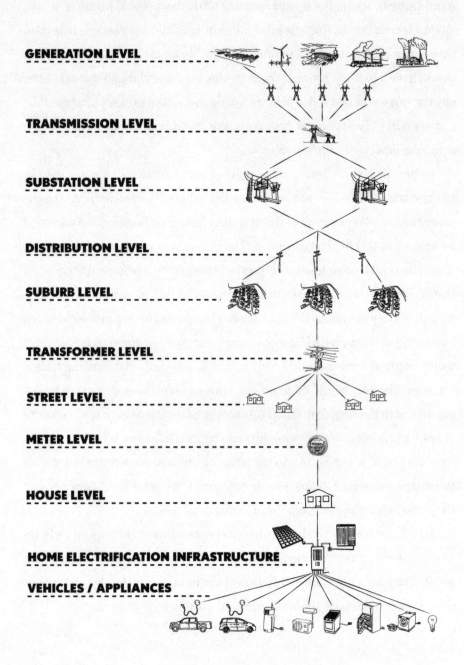

GENERATION LEVEL

TRANSMISSION LEVEL

SUBSTATION LEVEL

DISTRIBUTION LEVEL

SUBURB LEVEL

TRANSFORMER LEVEL

STREET LEVEL

METER LEVEL

HOUSE LEVEL

HOME ELECTRIFICATION INFRASTRUCTURE

VEHICLES / APPLIANCES

Lower bills for everyone

Electrification in the 2020s means lower bills and zero emissions. On the way there, pilot schemes will make the same offering available, thanks to modest subsidies from government. This investment will pay off for the rest of the community, because Australia will lead the world and create technologies for export.

What about renters, apartment dwellers and people who don't drive a car; how can they benefit from electrification? The financial structure of the pilots can be designed to include everyone in the community, even those who do not have a car or roof space for solar. The financial benefit that comes from replacing gas and grid electricity with cheap, locally produced solar can be shared. The way to make this work is that all households would sign up to an electricity retailer that is participating in the pilot. The retailer would offer reduced bills, regardless of how much rewiring a household is able to participate in. People such as renters in flats can still sign up to the pilot and receive cheaper electricity.

A well-designed pilot would have government provide up to half the cost of the purchase of all the devices, EVs, installation and project management. Much of the capital would be in the form of a loan, which would be repaid over several years. People could sign up to be part of the research phase of the pilot, which might go for two years, but they could keep the technology beyond that time. The revenue to pay back the loans will come from household electricity bills. Bundling all this income together via a single retailer will give a bank the confidence to lend against that annuity. For the first pilots, it is likely that concessional finance would be required, from the Clean Energy Finance Corporation (CEFC) or other government lenders.

The earliest projects of this kind were in Europe more than 20 years ago, when small groups of houses were offered primitive solar systems to augment the electricity purchased from the grid. One of the unexpected outcomes of the trial was that the community became more interested in energy: where it comes from, how it works and how they used it. That is something I can heartily endorse! And if electrification pilots provide people with a crash course in systems thinking and innovation, wouldn't that be wonderful?

Beating the brownfield blues

The building sector is already building electrification into housing projects. These 'greenfield' developments are easy to electrify because they are being built from the ground up. The hardest nut to crack will be electrifying established communities in which gas is used for cooking, hot water and space heating. The challenges will be both technical and cultural.

It is easier to replace an inefficient heat pump or electric stove with a better version than it is to replace gas appliances with electric ones. Replacing gas with electricity involves two trades (a plumber and an electrician), removing pipes, adding new conduit and upgrading the switchboard.

There is also the cultural obstacle. People who are used to gas can find it hard to make the switch. The solution is not to mandate that people cancel their gas, but to nudge them to consider the cleaner, cheaper alternative. The Ginninderry development in the ACT is a greenfield precinct that is going electric from day one. The developers organised for celebrity chefs to do cooking demonstrations on induction stoves to promote the technology to potential residents. The first households in a pilot zone to switch will become ambassadors for their neighbours. The best form of persuasion is social proof: seeing the change and hearing the message from people you already know and trust.

Government's role in funding energy innovation

Strategic shifts in technology have generally been funded by governments. The clean-energy revolution in our cities and towns will mostly be driven by consumers and private firms over the second half of the 2020s. But before the market really lets rip and the big money moves in, government has a critical role to play. Governments must turbo-charge the innovation processes and prove the electrification model to consumers, financiers and politicians.

Our national government has more power and money than the states, so it should fund the biggest and most expensive electrification pilots. The federal government created the $10 billion Clean Energy Finance Corporation and multibillion dollar Australian Renewable Energy Agency. These are the experts in clean energy innovation and have a track record of success.

The Australian Renewable Energy Agency (ARENA) comes first in the innovation process. It is a precommercial phase funder. It provides grants to technologies that have been proven to be technically capable as a concept or in the lab or at a small scale but need to be demonstrated in the field. Since it commenced funding in 2012, it has granted $1.77 billion to 602 renewable energy projects. One third of these were for solar PV.

The Clean Energy Finance Corporation (CEFC) is a green bank; it provides loans to deploy clean energy technologies that are commercially viable economically but have not been funded yet by banks. It 'crowds in' commercial investment funds, to join with government funds lent at concessional rates. Since it began in 2012, it has committed $9.5 billion to 220 big projects.

The CEFC and ARENA are already funding projects that include many of the technologies and challenges involved in electrification. Over the next few years, these agencies will drive much of the electrification innovation in Australia and this is likely to involve multiple pilots. The federal government can also support pilots directly through grants or loans from the Department of Industry, Science, Energy and Resources. Funding can also come indirectly via the CSIRO and universities, although these will tend to focus on earlier-phase research and development.

States and territories and even local government can help support pilots. States have their own clean energy innovation funding. New South

Wales, for example, has $75 million for pilots in its Net Zero Industry and Innovation Program. Local governments do not have the revenue base to provide large grants for technology, but they still have a critical role. The local level of government is closest to citizens and intimately involved with their communities. The City of Bendigo in Victoria is taking its community on a journey to reach zero emissions by 2030, through the Greater Bendigo Climate Collaboration. This would be a perfect starting point for a pilot of full electrification.

A race to the top

Australia is a great place to run electrification pilots because it is such a big country. We can test the technology in climate conditions that can match any place on the planet. A range of pilots in different climates and varying types of built environments would prove electrification works anywhere, from the edge of the grid in Morocco to the ski villages of Japan, from modern suburbs in Boston to the medieval heart of ancient cities like Budapest.

Australia could sell the smarts to the world by doing the innovation at commercial deployment scale. Communities around Australia could work with their local governments and state and federal members of parliament to put in proposals to host pilots. A genuine grassroots movement of clean energy citizens can lead governments to fund the innovation. The outputs from pilot schemes can become products and services we sell to the world. Innovation will be necessary, from the hardware to the software to the 'wetware' or culture and business models. The devices are all proven, but there remains some clever thinking to be done in the energy management systems that tie it all together.

Software will also be critical. Everyone should be able to readily find out how much money they are saving and the emissions they have avoided by electrifying their castles and cars. This knowledge is power and it should be only an app away. There will be important innovations in the business processes. The processes have to become easy for consumers to understand and manage. People need to be in control. The economic spoils should be shared fairly with consumers and not all sucked up in corporate profits for new or old energy companies.

If Australia can quickly jumpstart a range of pilots around the country, we can lead this revolution. A range of different approaches will reduce risk. It will put more prototypes in the field and result in faster innovation. The federal government can also play an important convening role, bringing stakeholders together around pilots. We need to retrain tradespeople to rewire houses. Banks need to get the confidence to finance the transition. The green tape barriers need to be cut away.

Grid neutrality

The rules of the electricity market will decide how quickly Australia can electrify and whether the consumer reaps their fair share of the financial windfall. Electricity market design and regulation is incredibly complex, so I like to explain what is at stake using a concept of fairness that people might be more familiar with from the internet.

One of the fundamental principles of a free and open internet is that your internet provider is not allowed to discriminate against you. All information traffic must be treated equally. The energy equivalent is 'grid neutrality', the idea that households, business consumers and generators operate as equals. All energy should be transported equally across our grid, whether it comes from your rooftop solar, your EV battery or a dirty old coal power station in the La Trobe Valley. Grid neutrality would lower the cost of energy. It would also improve reliability. It would encourage batteries and inverter-based solar and wind to provide the security services that coal and gas are paid for.

You might think this sounds so obvious that you wonder why I bring it up. Electricity law in Australia is biased against clean energy; it is not a neutral grid at all. Even more alarmingly, some of the rules of the electricity market are becoming more biased against renewables. For example, the energy regulator proposed a rule that would effectively double-charge batteries for doing what they are designed to do. A battery is not a generator of energy; it is a store for energy already generated. The idea was that batteries should pay for the use of the grid to store energy already generated and then pay again when they dispatch the energy out to the grid when it is required.

This is crazy. Without storage, there is no way that Australia can decarbonise. Old rules in our market are biased to coal, but it is inexplicable to see new rule changes that would lock in coal as our forever friend, just when it's time to move on to new, cleaner energy. The Glasgow Climate Conference made history by calling out coal as the cause of so much of the climate problem, and for it to be 'phased down'. Why on earth would our lawmakers want to protect coal from competition, just when the world has agreed it is time to switch it off? The reason, unsurprisingly, is money and influence.

Electricity might seem like a Wild West version of capitalism with the hundreds of retail offers available to you, but in truth it is a market entirely constructed by government. There is a single set of poles and wires going to your house. Electricity distribution is the textbook example of a monopoly licensed by government. Governments decide which generators are built and where, who owns the transmission and distribution networks, how much consumers pay to take power from the grid and how much generators (or batteries) pay to send power to the grid.

This means that the incumbents who dominate the market have a huge economic interest in making rules that protect their profits. They spend millions on consultants and lawyers and regulatory experts putting in submissions and lobbying. Their objective is to stop clean energy and households competing against their investments. Small-scale renewable energy is also being subjected to what critics have called the 'solar tax'. This rule will mean that households have to pay to use the grid when they export energy that other consumers need. Coal generators, meanwhile, don't have to pay this charge.

Last year the rule-makers made a big reform that should be great for electrification. It allows consumers to be paid to reduce their demand when this is cheaper than paying generators to increase supply. That would lower the wholesale price paid by all users of energy, so it benefits everyone, not just the consumers who sell their 'demand response' into the market. The big coal generator-retailer companies lobbied to ensure that consumers would not be free to sign up to a contract to do demand response. They also lobbied to prevent households from participating in the new market. Only big manufacturers and commercial firms are even allowed to participate in the demand-response market.

Can you guess who benefits from all these distortions of the market?

Solar owners have been offered very generous feed-in tariff payments at times. These schemes were often designed quickly by politicians wanting to win votes at an upcoming election. The rates have changed many times and vary across the country. This 'solar roller-coaster' has led to booms and busts and damaged the solar industry. Clean energy advocates sometimes ask for time-of-use pricing as a better solution. This means charging more when demand is high and less when supply is plentiful. This is a fairly crude way to steer behaviour. Most people aren't sitting around at home waiting for a price signal to tell them when to turn on the TV or do a load of washing or how to cool the room where the baby is sleeping. It would be far simpler to start with the universal principle of neutrality: households and big generators would be treated equally. All players in the market would be allowed to buy and sell from each other without limit. This is efficient in terms of both energy and economics, watts and dollars.

Nobody expects citizens to take on the coal generators in the complex regulatory debates. But you can tell your local member of parliament and everyone who follows you on Twitter that you're optimistic as hell and not going to take it anymore.

Don't hate the energy industry; be the energy industry. Work with your neighbours to get your local council to form an electrification pilot committee and to ask for funding from the state and federal government. If you are finishing school, like an argument and are wondering what to do in the future, consider a role as a clean-energy lawyer. We need everyone on board, to switch to clean energy as fast and as fairly as we can. Many of the smartest young people I come across are doing this already. They could be maximising their income as merchant bankers or commercial lawyers, but they are working in clean energy companies or for governments, accelerating the clean energy transition.

Green buildings

If we want everyone in Australia to be living the green dream by 2030 – zero energy emissions and lower bills – we must make our governments change the rules for buildings now. Every house or apartment block that is built with gas appliances is money down the drain and needless emissions in the air.

Our building stock is a massive investment that is built to last decades. We need planning laws to guide developers and owner-builders to construct smart, efficient buildings for the next generations. Electrification requires the right rules for the grid, but it also needs good planning rules for homes, commercial buildings and the form of towns and cities.

Where possible, all new houses should be built with enough solar on the roof to power the home and cars. Our switchboards shouldn't just be dumb fuses, but able to manage the solar and the vehicle charging and balance the heavy loads in the home and do their part in demand response, voltage management and grid resilience. There should be electric chargers in the garage or carport and room for a battery cabinet. Space and water heating must be efficient heat pumps and we need electric induction stoves.

The federal government runs the Nationwide House Energy Rating Scheme (NatHERS), which provides a standard methodology for assessing how thermally efficient a home is. It is a scale from zero to ten, where ten stars means the house does not need almost any space heating or cooling to be comfortable all year round. In 2010 the Building Code of Australia was updated to require new homes to be six stars or better. This is well behind international benchmarks, and there is a push to raise the minimum to seven stars, which would translate into a reduction in energy use of 24 per cent.[39] Increasing the thermal efficiency would mean an increased build cost but financing this through a mortgage would be less than the savings in energy costs.

Luckily, many owner-builders and developers are moving ahead of building standards. It turns out that people want to be healthier and more comfortable and live in better-designed houses. The Green Building Council of Australia runs a voluntary 'Green Stars' rating system. This encourages full electrification and also measures the 'healthiness' of the house (ventilation, comfort) and climate resilience.

There are many ways that buildings can be cooler for occupants and make neighbourhoods cooler. Light-coloured roofs keep buildings cooler. Vertical gardens and shade trees help houses handle heatwaves. Passive solar design means orienting and structuring a building to help it keep out summer heat and retain winter warmth. Half the energy consumed by a building over its lifetime is embodied in its construction. This means we need to find lower-carbon materials and make them long-lived. Australian homes built from wood with steel roofing materials are quite good and can actually be a carbon sink. This isn't true for expansive cement and concrete work, which is one of the highest CO_2-emitting products. Embedding carbon in our building materials rather than making our homes sources of carbon is a great idea. We need to make homes well and improve the quality of the industry and building stock – and the best way to lower the embodied energy burden is to increase the lifespan so the burden is amortised over a longer period.

Community batteries are mid-sized battery installations that are shared by all residences and businesses. This is particularly important in high-density communities, where there is less roof space per resident and therefore less solar potential.

Electric vehicles are storage on wheels and they should be used to back up the grid. This means having chargers out in the community so that cars can be charged at work in the middle of the day, when there is excess solar production. These rolling reservoirs can then be discharged when parked at home during the evening. This matches the peak of domestic consumption around 7 p.m. It is when the sunshine is waning and stoves, ovens, space heating/cooling, dishwashers, fridges, TVs and washing machines are getting their workout. From this year, all new buildings in England will have to include EV charge points. It is projected this will result in around 145,000 new charging points being installed each year.

Cities also need to be designed from the top down. This means better mass transit (trains, trams and electric buses), and bike paths that are not just for weekend sightseeing. It means a higher urban canopy cover, with more parks and trees along roads. It also means waterways and water conservation.

If we use pilot programs to work out how to electrify our existing building stock, establish stronger standards to ensure new buildings are electric and build community EV chargers and batteries, then our cities and towns will be well on the way to a cleaner, better future.

Rebates, retraining, supply chains and big finance do the heavy lifting

Once pilot schemes prove electrification in the early years of this decade, the pace of rewiring can really pick up. Our modelling envisages that, mid-decade, we could be retrofitting electric devices into 10 per cent of households each year. This is a massive logistical exercise as well as a considerable financial outlay. At its peak, we predict a cumulative cost of $12 billion.

There are four policy measures which need to be implemented in lockstep to allow that pace of rewiring by around 2025. Government rebates are a form of subsidy that are quite pure and can go directly to the Australian consumer. For low-income houses, rebates are an ideal mechanism to bring down the upfront costs of the clean energy solutions that enable that household to decarbonise. This will ensure the whole community benefits at the same time. Rebates should be available on solar, heat pumps, EVs, household efficiency and electrification upgrades, and more. The best rebates are applied at the point of sale, when the item is being purchased. Mail-in or other more complicated methods of getting the reimbursement add unnecessary friction.

Retraining and associate labour force support is vital if we are to be able to install 50 million electric machines in our households and driveways in a decade. As with rebates, this is best led by the federal government and would mean working closely with unions and the technical trade sector. Certification and training programs analogous to what we did for rooftop solar, only for HVAC, batteries, EVs and smart switchboards, would be a world-leading start.

Supply chains are a potential limiting factor and the federal government should also take the lead here. Australia needs to ensure we have access to the best-in-class batteries, EVs and appliances. The government should forge relationships with key manufacturers and manufacturing countries and build them into the plan. Finance can be encouraged by concessional finance and loan guarantees from the CEFC and other government mechanisms. Government support will trail off as private money flows into the mortgage and other financial products which will pay for electrification.

The '20s are roaring already

By the second half of the 2020s, car manufacturers will all be dumping ICEs for EVs. Banks will offer battery, solar and electric appliance packages in mortgages. Real-estate agents – everyone's favourite people – will tell you to avoid the house fuelled by fossils, to save money and live cleaner in an all-electric home. Our communities will have shared batteries and EV chargers.

That huge pool of power will absorb excess solar in the middle of the day and give it back to us when our work is done and domestic life amps up. Year on year, our suburbs and towns will become cleaner, cheaper places and quieter. Those damn leaves are louder than my electric leaf blower!

We will all be able to sleep at night knowing our domestic energy consumption is climate-friendly. As I have argued in this chapter, the next few years are critical and could position Australia as the world leader in electrification. But before we can rewire our suburbs and towns, we have to short-circuit the influence of dirty energy over the electricity laws and system.

Everyone can be an expert in the kind of safe future they want for their children. Everyone can work in their community to lobby their local council to become the champion of rewiring in your area. The states are already leading in renewable energy and EV policy. If we put together the best ideas and rules and visions from each state and territory, this best-of-Australia can be a recipe for the rest of the world. The federal government can lead a plethora of pilots across the country, to prove out the model. Out of these innovation incubators will come new technologies for energy management, new models for marketing the transformation, and new ways to share the benefits fairly.

Chapter 10
Financing fossil freedom

- Long-term finance will be critical for homes and businesses to afford the upfront costs of electrification, which will ultimately save them money in the long run.

- It is vital we make sure affordable finance is available for low-income households so that all Australians can benefit from our electrified future.

- Given we can forecast this electric future saving Australians money, state and federal governments should be designing financing and tax policy around it.

Someone's gotta pay for it.

In 2018, when I was only a bit younger and a bit more naive, I sought a meeting with David Gonski. My father knew of him through their mutual history with the University of New South Wales. This is not why I was seeking out David. I was seeking him out because I was interested in finance, and I was interested in his role as board chairman of ANZ.

I still have the presentation I pitched him with in early 2018. 'The audacious plan for a net positive energy Australian housing stock: A new financial product?' I laid out the argument, largely repeated in this book, that Australia would be the first country in the world with the right conditions for highly profitable financial products for Australian citizens to finance retrofits and upgrades to their cars and houses that would save them money and get them to zero emissions. I laid out the household economics. I was seeking his ideas and a big-picture conversation. I think he was merely being polite in taking the call and didn't really understand why he was there.

After some explanation of how the physics worked, I think he sort of believed the larger argument that renewables-based, demand-side electrification was the answer. Now I wanted to know how to keep the interest rates low for financing important things. I wanted to know how to get the banks to make these financial products exist. At a 7% interest rate for ten years, you double the cost of something, as expressed simply in Figure 10.1. At a 2% interest rate, the cost only increases 20%. Interest rates matter. A million

10.1: Effect of interest rates on the cost of a high-capital item over a 20-year financing period. A 3–4% loan roughly doubles the cost; a 9–10% loan increase it six- or seven-fold.

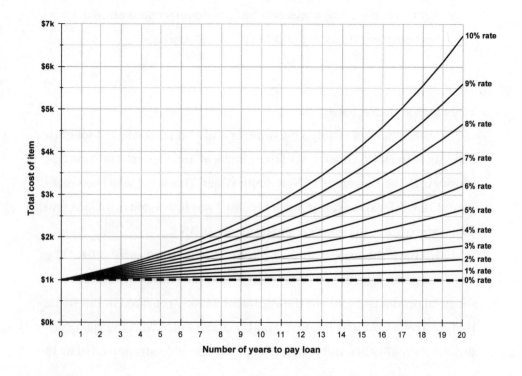

conversations in a million golf clubs and bowling clubs are testament to that. Interest rates are hugely influential on Australian homes.

Our homes may finance at 2–5%, but our cars finance at higher rates. I was, and am, interested in how to make the grand project of decarbonisation as inexpensive and as beneficial as possible for the Australian people. There is precedent for caring about this interest rate, and about financing consumer goods. Australian lifestyles are built on loans. The car loan and mortgage were both 20th-century American innovations and the modern world would be unrecognisable without them. They are how the bulk of the population afford big-ticket capital items.

Creating a climate loan in response to the climate crisis has clear historical precedent. Henry Ford wouldn't allow his cars to be purchased on debt because of his religious beliefs. General Motors' Alfred P. Sloan recognised the market opportunity of making cars affordable to the masses by inventing auto-financing. This American innovation was the precedent for the modern home loan. The modern mortgage market was shaped by the US government's intervention in another time of crisis: the Great Depression. During the Depression, property values plummeted, and about 10% of all homeowners faced foreclosure. The government stepped in as part of Roosevelt's New Deal, when Congress passed the *Home Owners' Loan Act* of 1933, which created the Home Owners Loan Corporation (HOLC) to provide low-interest loans for families at risk of default. As a result, hundreds of thousands of homeowners were able to pay off mortgages, and the program actually turned a slight profit, defying expectations of massive loss of taxpayer money. This program gave rise to the first modern mortgage provider, Fannie Mae, in 1936, and then to Freddie Mac in 1968. This led to the lowest-cost debt pool and largest capital market the world has ever seen. Australian banking followed in these footsteps. The 20th century, the suburbs and the proliferation of cars wouldn't have happened without it. We can debate the good and bad implications of cheap consumer financing, but it is hard to dispute that it shaped the world. I think it had more impact on the 20th century than any other innovation.

It was in the spirit of thinking big-picture about the interactions between governments and banks, and about fiscal policy writ very large, that I had approached David. After about half an hour of talking, I think he didn't think I was insane. After another 15 minutes, he suggested it might even work. After 15 more, he said that I shouldn't go into the banking business: it's competitive and a race to the bottom.

Katherine McConnell of Brighte.com.au, a startup interested in financing Australian home improvements, particularly green ones, is now creating the financial instruments to allow homes to purchase these climate upgrades while also helping the contractor and tradespeople network with financing to cover their cashflow. It's working and her business is growing exponentially.

Not much came out of that conversation with David Gonski for me, but when I returned to Australia in 2020 I was contacted by Macquarie Bank. It had a new division of people focused on retail consumer financing for all the components I think we need to address our domestic emissions. Electric vehicles, rooftop solar, electrified heat and kitchens, and efficiency retrofits. Low-cost financing against existing mortgages. The bankers I talked to were kind enough to credit me with some small influence on their product launch because I'd been talking so much about the importance of consumer finance in making climate solutions affordable. Who knows if I had any influence, and it doesn't really matter, but a lot of Australians can now use Macquarie or Commonwealth Bank or products from Brighte to help upgrade their castles. The federal government, through various fiscal options, including guaranteeing the financing for certain groups, can help to keep these interest rates low and, critically, to expand the group of people who have access to these credit instruments. It could even act as a bank for this vital transformation. The point is, Australia is also beginning to lead the world on the financing of the electrify-everything revolution, and we could do more, and model good behaviour for the world to follow.

Another aspect of the good behaviour is figuring out how to finance everyone. It's not a fix if only the wealthy can afford it. This isn't some bleeding-heart socialist observation. All the emissions have to stop, not just the rich ones. A good credit score is a prerequisite for getting a loan. So are things like a steady pay cheque, and even that may not be enough.

This became abundantly obvious to me in conversations with my cousin. He is an HVAC technician who installs air-conditioners and refrigerators from Sydney to Narooma. He's exactly the type of person we need to retrofit and electrify Australia. He is one of the tradies out there building the world we all live in, and he's installing the heat pumps that are a critical component of it. My cousin was going for a loan but was struggling to get one because he is self-employed, which is deemed less reliable by the banks than a pay cheque from an employer. I was getting a small loan at the same time to do some renovations and was getting offered interest rates under 2%. He was getting offered rates about 5%. We shouldn't be handicapping the very workforce we need to electrify the future in their affording that future.

My cousin ultimately got a loan and moved in his young family and is already electrifying his new house. He looks forward to getting an electric ute for his next work truck. He's in the lucky camp on the right side of finance, even if it is hard for him to get a loan. For many people it is impossible. Their credit history, age, marital situation, work status, criminal record – all these things can work against them. I won't be prescriptive here, but it is important to contemplate all forms of access to this future we need everyone to embody. We'll get a better climate if we don't have to wait until everyone can afford the second-hand version of it.

One possible solution is already under trial in the US. There (and it isn't too different in Australia), 36% of households are rented. There is less of an incentive for renters to invest in home efficiency or electrification. Also in the US, 40% of households cannot cover an unexpected $400 expense.[40] One third of US households face energy insecurity, which means uncertainty about paying their bills.[41] We know that water heaters and air-conditioners are often bought under financial duress. The climate-friendly models typically cost more upfront and so don't get purchased, even if they would lead to savings over time. This group doesn't have traditional credit access, and even rebates are difficult if they are not applied at the point of purchase. What has worked is Inclusive Utility Investment. My friend Holmes Hummel and colleagues at Clean Energy Works are big proponents of investment instruments that do not disqualify on the basis of credit score, renter status or income. Utilities in this model use their special regulatory status as providers of essential services to provide financing to any household, blind to the occupants' credit history. The costs are absorbed on the monthly bills and through tariffs. The model program for this method is 'Pay As You Save' and 20 utilities in ten US states have so far spent $45 million at 6000 locations with upgrades funded this way.

The aftermath of the Global Financial Crisis, in combination with COVID's paralysing effect on the global economy, has interest rates as low as they've ever been. In Europe they are sometimes negative. We are at a great moment in history for the low-interest financing and other fiscal methods that can help everyone afford this vital energy transition to a zero-emissions world. The world has solved problems like this before. Economists have been instrumental in making things like this work before. It's time to think bigger than small-minded economics on this issue.

Chapter 11
So long, and don't kill all the fish

- Climate change isn't the only environmental emergency; there's also pollution, plastics and the biodiversity crisis.

- We should more actively consider solutions such as mechanisms to promote smaller cars, more public transport and fewer roads.

The rosy picture I paint in this book isn't so naive as to suggest that we can't fuck it all up. We could solve climate change without addressing our plastic waste problems, our falling sperm count, the nitrogen problem and algal blooms born of modern agricultural practices. We could over-fish the oceans and collapse critical ecosystems. We could continue to divide terrestrial ecosystems with roadways that kill our precious critters; in the US, you are never more than 20 miles from a road.[42] I'd bet nearly every Australian has experienced the anguished thud of a roo, possum, wombat or lizard. I remember reading a story about an American soldier in Afghanistan who was taken aback by the damage that goats have done to the ecosystem there. Apparently he asked an elderly Afghan woman why the country had allowed goats to eat everything. The story goes that the elderly Afghan woman's response was, 'Why did you allow cars to eat everything in yours?'

We could get to zero emissions but still bury ourselves in things we don't need. I've provided a narrative of largely keeping our toys and our hobbies and our institutions and traditions – our way of life – while addressing climate change. I believe we could become better stewards of the world and our fellow creatures. But I don't want you to think I'm a naive techno-optimist. I can see how in solving this problem we might create the next one, unless we also learn to recycle our windmills and solar panels, unless we learn to do agriculture without the same levels of pesticides and fertilisers, unless we expand the wildlands of the world and give the other creatures of the planet a little more room to thrive.

Just to put it all in perspective, consider that all the world's animals weigh about 2 billion tonnes.[43] This is approximately the same weight as all the cars in the world. Only 167 million tonnes are mammals, more than half of which are our livestock. Only 70 million tonnes are wild mammals. That's right, cars outweigh wild critters by something like 30 to one. We should probably contemplate that as we make road and vehicle policies. As of late 2021, 1.25 million people had reportedly pre-ordered a Tesla Cybertruck. The Tesla cybertruck weighs 3000 kilograms. Those unbuilt Tesla trucks will weigh half as much as all the wild mammals on Earth. They'll drive (most likely carelessly, sometimes autonomously) over roads paved and unpaved, further disrupting non-human life on this planet. Some people see a shiny new electric truck; others will see a divided ecosystem and a dead elephant.

I'd like to register a moderating voice and suggest that we might be best served with some things smaller and some things less. The planet would be more verdant, and quieter, with more electric bicycles than electric cars. More trams than electric pick-up trucks. Certainly freeing up wildlands for wild animals by changing our diet would be great.

In 1949 the Japanese government made a new class of cars, the Kei class, that were limited to tiny (150 cc, or 0.15 L) engines, and had strict limits on height, width and length. These smaller vehicles were given special tax breaks designed to provide families with affordable independence and mobility. In the 1950s the engine sizes were upped to 360 cc, in the 1970s to 550 cc and in the 1990s to 660 cc. The maximum power was set at

47 kW. In 2013 Kei cars reached their peak of 40% of the Japanese market. Some models, like the Suzuki Jimny, even become popular in foreign markets like Australia.

In the rest of the world, vehicles doubled in size. The Land Rover of 2020 weighs double the 1960 Land Rover. The same is true for the Mini, the Fiat 500 – most cars, in fact. We traded up in size for more speed, more safety, more cup holders, but the cars didn't take us any further; in fact the speed of traffic in modern cities is falling because of congestion. Australia is likely about to import ungodly numbers of giant electric four-wheel drives from Tesla, Ford, Rivian and others. Our road rules, tax codes and import policies currently favour ever-bigger vehicles that use more space, more materials and more energy. We could instead do something like Japan and motivate a second class of vehicles for local streets that are smaller, cheaper and ultimately less damaging.

We are making a similar mistake with electric bicycles. Overanxious government types have limited the power of an electric bike to 200 W. With modern hub-drives these bikes are quite good, but not sufficient to effortlessly carry heavier loads up bigger hills. I've been building electric bicycles of all kinds, but mostly cargo styles, for two decades. They need about 500 W or maybe as much as 1000 W if you want to make them capable of carrying a few kids or the dog and some groceries. Subtly, Australia is preventing a cornucopia of lightweight electric vehicles, e-scooters, e-skateboards, e-bikes, e-mopeds, all of which would provide far cheaper, faster and more convenient local transportation options. Why? We've written the rules of the world for fossil-powered monster trucks.

These design and policy choices make a difference. As we move forward with this incredibly exciting energy transition, let's not forget to ask ourselves some hard questions about what we want the world to look like. Wouldn't we all be happier if it were safe for our children to walk or ride to school on garden paths and bike trails, rather than dodging 3000-kilogram bullets on 60 km/h streets?

More walkable cities, more designated wild spaces, more well-managed forestry. More of a lot of good things, but maybe what we need more than anything is more time. I lived in the US for 20 years. They have precious little vacation time and work too hard. Most families are stressed about work, time and money. Parents drive furiously between work, school, caregivers and their first and maybe second jobs. There is no time left for mending themselves, let alone the planet. Australia is slowly lurching down the same trajectory. If we had more time, would we demand such fast transport? Would we amble to the local shops more often and sit to have a cappuccino in a reusable cup instead of opting for disposable and takeaway so that we can make our next hurried appointment? With more time, would we garden more, or contribute to tree-growing efforts in our communities? If we are thinking about energy abundance and an abundant Australia, isn't one of the best uses of that abundance an abundance of time? Time to enjoy the natural world around us – to heal it, and to heal ourselves.

Chapter 12
An abundant Australia

- This revolution isn't just for city folk.

- Everyday Australians are already winning and we could help them win more.

My mate Fiona, who lives in New South Wales, reminds me that the switch to widespread electrification is happening already, filtering into our everyday lives, making them better, more resilient. She lives humbly on the South Coast, where I think she spends most of her time either advocating for positive change or sharpening her wit. Always good for a yarn, she tells the story of Jacko (not his real name), her neighbour in Moruya, who is emblematic of why life might just be about to get better for all of us.

On the last day of 2019, the Black Summer bushfires that had been burning for over a month reached the sea near Batemans Bay. The power was out for day – in some areas weeks.

Jacko is 74. He likes fishing and, after years of working as a dairy farmer and chicken sexer, among other things, he sold up and moved to the coast. Jacko likes to issue weather reports in his standard unit of measurement – the number of dogs blown off a chain. Jacko is both types of gold – B&H unfiltered and Nescafé.

During the fires, Jacko rigged up an elaborate watering system that sprayed a fine mist over his house. The problem was, everyone else had the same idea and the water pressure dropped precipitously. Jacko had a tank, and a pump, but not enough petrol to run the generator for 24 hours, until the immediate risk had passed. The petrol station in town had closed because the pumps run on mains power.

Jacko can live without power for a few days, no worries, but his freezers can't. Every few months Jacko goes up to his brother's farm near Albury and shoots feral goats and sometimes deer. He has three chest freezers stuffed full of the silliest ungulates in the Riverina.

During the hot nights that followed the fires, the freezers began to 'burp'. Jacko discovered that there are only so many times in a man's life that he can deal with a chest freezer belching dead roo. He cracked the shits and called his mate from the bowls club who had a new 'off grid' system.

Two weeks later, Jacko bought a Tesla Powerwall and installed a rooftop solar system. He's not interested in the 'bells and whistles' of the app on his phone, but he can see that the roof of his three-bedroom, double-brick home produces more power than he uses. Importantly, it means that his grandson can come over and use his nebuliser whenever the mains power goes out. He's changed a few things over, including adding an induction stovetop, which, aside from being easy to clean, means he's less likely to set the house on fire if he leaves the rhubarb on.

Ray (from bowls) has a Tesla car that he plugs into his system and uses as a battery. Jacko's hanging onto his diesel ute for another couple of years until someone makes an electric one because he's not having Blue, his enormous cattle dog, in the car with him. But he did buy an old electric golf cart so he can pop down to the wharf when his knee is playing up.

Growing up on the farm at Albury in the 1950s, Jacko's family was pretty self-sufficient, but it wasn't an easy life, living with an ammonia fridge and hand-pumping water for the house, garden and the odd bath. Now, he's got a veggie garden, water tanks and rooftop solar system that produces so much juice he can have a spa bath on the deck at night. He's not worried about power bills. He doesn't think twice about leaving the water pumps on for the veggie garden or running the small air-con unit in the front room on the

40-degree days when the grandkids come over to play Uno. His home is not just his castle, it's his kingdom.

Australia needs a good dose of hopium. Every mainstream party has driven a negative debate around climate change. Under duress, they will eventually 'admit' that this negative approach is a 'problem'. It's then framed as both a negative thing and a pesky cost that some dreary, whining greenie who doesn't even know how to start a tractor is forcing them to pay. But there is plenty of hope to go around. Australia is an abundant land. We just need to allow ourselves to tell, and hear, the stories about how doing things slightly differently (not shockingly differently) can make it all turn out OK, such that we can focus our attention back on the cricket/footy/surf/horses/fishing/bowls, or whatever is your Saturday-arvo thing.

For whatever strange reason, my father had Jaguars when I was growing up. They smelled like walnut, leather, pipe smoke and burnt engine oil. If you could design an aftershave for me, that's the smell I'd like it to suggest. I was no stranger to cars, boats, trucks and odd vehicles. One of my fondest memories is of driving with my parents to a movie with them in the last of Dad's Jags. It was raining and wipers squeaked over the front window, and from the backseat I saw the silhouette of my dad, one arm on the wheel and the other around my mother's shoulder. She had sidled over to him on the bench seat and her silver hair sparkled and caught the reflections of raindrops and oncoming headlights.

I know that cars aren't great for the planet so somehow I struggle forward as a climate hypocrite, trying to resolve my love for the machines and the moments. My wife and I are onto our third leased electric vehicle, having owned a Fiat 500e, a Chevy Bolt and a weird Nissan electric van based on the Leaf. That last one was perhaps the most practical vehicle I've ever owned, a bucket for dogs, sand and children that practically runs for free. I don't go much for new cars, so this year I electrified my 1957 Fiat Multipla and started in on my 1961 Lincoln Continental. The Fiat makes a VW beetle look big but nevertheless seats six and was the dominant taxicab in Italy in the '60s. It's hard to tell the front from the back. I'm retrofitting it with six oversized electric skateboard motors and a bunch of hobbyist parts, and it will easily get me around my neighbourhood and the kids to school. The Lincoln is a stunning car. It was released in 1961 as a rejection of the finned looks of '50s Detroit, streamlined and space-aged, perfect to become the presidential limousine of the first space-age president, John F. Kennedy. Ford has electrified the Mustang, and is famously electrifying the F-150, the most-produced ute in history. Because of this there are now aftermarket parts for retrofitting the classics. I was one of the first lucky people to get my hands on the Eluminator, Ford's 280 Hp electric-crate engine. It replaces not only the engine, but the transmission and the differential. That's one of the great things about electric cars – fewer parts, less failure, less oil, less grease. For far less than the cost of a new electric car, my Lincoln Continental will be fully electrified by the time this book comes out. It will mostly sit in the garage, I suspect, where it will be the backup battery for the house. That's how I sold the project to my wife – my vintage car isn't a 'toy' anymore, it's a critical piece of household resiliency and part of the national infrastructure. My father borrows my electric Nissan van whenever he can. It's the only way he can use all the solar being produced on his roof. He bores his golf mates with reports of how many dollars he generates or saves every day. Like Jacko, he often uses his electric golf cart to run an errand or save his knees. At 83, he is resisting getting an electric car for himself but is enjoying being part of the transition.

My sister lives on the northern beaches and this year installed her second solar system and a house battery, and bought herself a Hyundai Kona electric. Her stovetop is now induction and the gas water heater has been replaced with a heat pump. She's a single mum and has found that all these decisions make economic sense when she plans her life and the independence she wants. She works in sustainability and teaches the next generation of kids about how to think bigger and more entrepreneurially about climate solutions. She's walking the walk.

I have an old mate, Colin – I first him met when I was a grad student at the University of Sydney and I was helping teach a class he was taking in engineering. He lives in the Southern Tablelands now and is building his own electric aircraft designs, which will soon be available as kits. The carbon-fibre two-seater can easily fly a few hundred kilometres and should be safer and far more reliable than similar planes that run on petrol. It won't be long before our fly-in, fly-out workers are flying in and out to lithium mining and smelting operations in electric aircraft built by people like Colin.

My cousin Harley runs a very normal Australian business. He installs HVAC systems, specialising in refrigerators for small businesses. He drives the truck and does the installation work while his wife, Tiff, does the books; they service the south coast, from Narooma to the western suburbs of Sydney. Like many tradies, he will benefit hugely from the electrification of our homes and businesses and from this great energy transition. His next work truck will be electric. He's got solar on his new house, providing clean energy to his two young kids. His children will never have to experience the respiratory challenges of growing up in a house heated with natural gas because he's already gone electric.

The future is already happening in Australia. We are three to five years ahead of the US on this journey of electrification (except for our abysmal record on electric vehicles). We are ahead of Europe. We are the logical country to lead the world and show everyone that the future will be awesome, electrified and abundant.

To achieve this, we can't afford to let up the pressure on our politicians and the sooty hands of big business. They need to hear the good news stories and imagine what success looks like. Everyone does. Everyone wants to be on the winning team. We may seem like the climate underdog, given the last 20 years of crappy Australian federal government climate policy, but the underdog here can come good and bring it home with a strong finish this decade, winning the climate future we all want for Jacko and for Greta, for ourselves and for our children.

Rewire Australia.

Electrify everything.

Appendix:
Scales of energy use

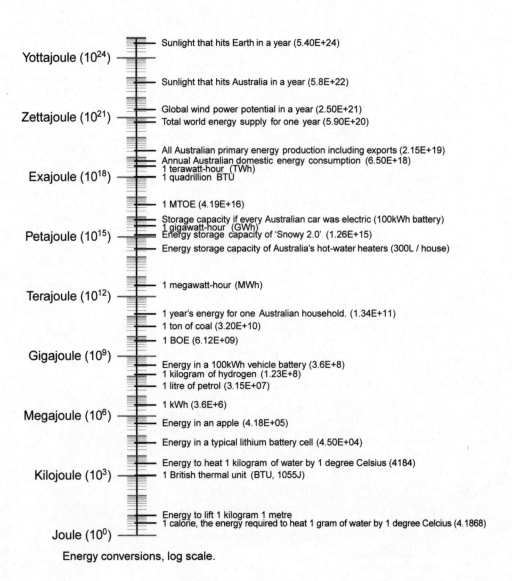

Yottajoule (10^{24}) — Sunlight that hits Earth in a year (5.40E+24)

Sunlight that hits Australia in a year (5.8E+22)

Zettajoule (10^{21}) — Global wind power potential in a year (2.50E+21)
Total world energy supply for one year (5.90E+20)

All Australian primary energy production including exports (2.15E+19)
Annual Australian domestic energy consumption (6.50E+18)
1 terawatt-hour (TWh)
Exajoule (10^{18}) — 1 quadrillion BTU

1 MTOE (4.19E+16)
Storage capacity if every Australian car was electric (100kWh battery)
1 gigawatt-hour (GWh)
Petajoule (10^{15}) — Energy storage capacity of 'Snowy 2.0' (1.26E+15)
Energy storage capacity of Australia's hot-water heaters (300L / house)

1 megawatt-hour (MWh)
Terajoule (10^{12}) —
1 year's energy for one Australian household. (1.34E+11)
1 ton of coal (3.20E+10)
1 BOE (6.12E+09)
Gigajoule (10^{9}) —
Energy in a 100kWh vehicle battery (3.6E+8)
1 kilogram of hydrogen (1.23E+8)
1 litre of petrol (3.15E+07)

1 kWh (3.6E+6)
Megajoule (10^{6}) — Energy in an apple (4.18E+05)

Energy in a typical lithium battery cell (4.50E+04)

Energy to heat 1 kilogram of water by 1 degree Celsius (4184)
Kilojoule (10^{3}) — 1 British thermal unit (BTU, 1055J)

Energy to lift 1 kilogram 1 metre
1 calorie, the energy required to heat 1 gram of water by 1 degree Celcius (4.1868)
Joule (10^{0}) —

Energy conversions, log scale.

Acknowledgements

There is no way this book would have been completed without Josh Ellison, who embodies the spirit, energy and urgency that we need from young Australians to address this climate issue. He made graphs and images and wrote text and organised footnotes and made suggestions and improved the book in every way. He really deserves his name on the cover of the book.

Sam Calisch and Laura Fraser were my collaborators on *Electrify!* and both came in here to scrub and polish. Laura taught me how to add a bit of love story to largely technical topics. Although I was a mentor to Sam, who was my intern way back in the day, he is now a favourite friend and peer.

Thanks to Danny Kennedy for encouragement and friendship and wisdom and walks. He was a solar pioneer and is now pioneering clean energy incubators across the world. The world needs more Danny.

Thanks to Dan Cass. I met Dan through Danny. It is hard to describe Dan's superpower. Maybe it would be called 'complete lack of tolerance for bullshitdom, served with a smile'. He's not just productive; he makes everyone around him move at lightning speed. Dan wrote most of the policy section. I love your work, Dan.

Thanks to Eytan Lenko. Eytan has been fighting the zero-emissions fight in Australia for what seems like decades already. His organisation, Beyond Zero Emissions, is a leading light nationally and globally.

Thanks to Gabrielle Kuiper, who dropped some bread crumbs and is an original gangster for a sustainable Australia.

Thanks to Martin Green, who really should be knighted for his contributions to solar.

Thanks to Matt Kean, who has the courage to genuinely lead on climate.

Thanks to Emma and Dave Pocock, who along with other brave Australians are putting their creativity, their strength and all their energies to this critical transition. Good luck, David, in the upcoming election and in becoming a federal voice for climate policy sanity.

Thanks to Christine Milne for the best breakdown of Australian climate politics I have ever heard, and for all of her contributions in demanding a better world and representing not just the people who don't get heard enough, but the trees and critters that barely get heard at all.

Thanks to Mike Cannon-Brookes. He once described himself as someone who fell out of the lucky tree a thousand times and never broke a bone, and that's why he feels compelled to put his reputation and his money behind solving climate change. I feel like I fell out of the same lucky tree by earning his friendship.

Thanks to Ben Oquist and the Australia Institute, who took a punt on supporting the creation of Rewiring Australia to agitate for proactive electrification policy in Australia. Ben understands political dynamics in Australia such that engineers like me don't have to.

Thanks to Curtis and Louise for the sailing and the hang-gliding. Everyone needs to know why it is so damned great to be outdoors in this awesome country.

Thanks to Tim Flannery, whom I'm now lucky enough to call a neighbour. I read and laughed along with your books long before I knew you and wasn't at all surprised to meet a fabulous human being behind all those fabulous stories.

Thanks to Chris Feik and his team at Black Inc. I still can't believe you believed we'd write this book in under two months and get it onto shelves with so much urgency. Bloody marvellous.

Thanks to Fiona Whitelaw. I've never met a tongue more acerbic, a heart more golden and a mind more hilarious. Keep being inspirational.

Thanks to my mother, Pamela, and my father, Ross, and my sister Selena, who apart from tolerating me and shaping me, all read early manuscripts and provided edits and suggestions. And for fuck's sake, Mum, I didn't remove all of the swear words – they were carefully chosen for emphasis.

Endnotes

1 D. Tong, Q. Zhang, Y. Zheng et al., 'Committed emissions from existing energy infrastructure jeopardize 1.5°C climate target', *Nature* 572 (2019): 373–77.

2 H. Rosling, 'The rapid growth of the world population: when will it slow down?' Gapminder, www.gapminder.org

3 H. Rosling et al. Factfulness: *Ten Reasons We're Wrong About the World – and Why Things Are Better Than You Think*. Flatiron Books, 2018.

4 S. Arrhenius, 'On the influence of carbonic acid in the air upon the temperature of the ground', *Philosophical Magazine and Journal of Science* 5:4, April 1896: 237–76.

5 BARES 2021, *Snapshot of Australian Agriculture 2021*, Australian Bureau of Agricultural and Resource Economics and Sciences, doi 10.25814/rxjx-3g23

6 Department of Industry, Science, Energy and Resources, *Australian Energy Update 2020*, September 2020.

7 P. Yanguas et al., 'Evaluating the significance of Australia's global fossil fuel carbon footprint', Climate Analytics, July 2019, climateanalytics.org

8 R.D. Kinley et al. 'Mitigating the carbon footprint and improving productivity of ruminant livestock agriculture using a red seaweed', *Journal of Cleaner Production* 259 (2020), doi 10.1016/j.jclepro.2020.120836.

9 Bureau of Infrastructure, Transport and Regional Economics (BITRE), *Growth in the Australian Road System BITRE*, 2017.

10 A. Blakers, B. Lu and M. Stocks, '100% renewable electricity in Australia', *Energy* 133 (August 2017): 471–82.

11 M. Roberts et al., *How Much Rooftop Solar Can Be Installed in Australia?* Report for the Clean Energy Finance Corporation and the Property Council of Australia. Sydney: University of Technology Sydney, 2019.

12 Australian Energy Market Commission, 'Residential electricity price trends, 2020 data', www.aemc.gov.au

13 M.Z. Jacobson and C.L. Archer, 'Saturation wind power potential and its implications for wind energy', *PNAS* 109:39, 25 September 2012.

14 Jacobson et al., '100% clean and renewable wind, water and sunlight all-sector energy roadmaps for 139 countries of the world', *Joule* 1:1 (6 September 2017), doi 10.1016/j.joule.2017.07.005.

15 International Hydropower Association, *2020 Hydropower Status Report*, hydropower.org

16 S. Mallapaty, 'China prepares to test thorium-fuelled nuclear reactor', *Nature* 597 (9 September 2021): 311–12.

17 D. Wallace-Wells, 'Ten million a year', *London Review of Books* 43:23, 2 December 2021.

18   Australian Renewable Energy Agency (ARENA), *Australia's Bioenergy Roadmap*, November 2021; and T. Nugent, *Australian Biomass for Bioenergy Assessment 2015–2021, Final Report*, April 2021, arena.gov.au

19   Department of Energy, *2016 Billion-ton Report: Advancing Domestic Resources for a Thriving Bioeconomy*, July 2016, energy.gov

20   T.S. Brinsmead, J. Hayward and P. Graham, *Australian Electricity Market Analysis Report to 2020 and 2030*, CSIRO Report EP141067, 2014.

21   N. Kittner et al. 'Energy storage deployment and innovation for the clean energy transition', *Natural Energy* 2 (2017); and L. Goldie-Scot, *A Behind the Scenes Take on Lithium-ion Battery Prices*, BloombergNEF, 2019.

22   G. Parkinson, 'Australia could aim for 700 per cent renewables, Arena boss', Renew Economy, 8 October 2019, reneweconomy.com.au

23   Energy Networks Association, *Fact Sheet: Electricity Prices and Network Costs*, energynetworks.com.au; and Australian Competition and Consumer Commission, *Retail Electricity Pricing Inquiry: Preliminary Report*, Commonwealth of Australia, 2017.

24   For example, see N. Eyre, 'From using heat to using work: reconceptualising the zero-carbon energy transition', *Energy Efficiency* 14:77 (2021), doi: 10.1007/s12053-021-09982-9

25   T.P. Wright, 'Factors affecting the cost of airplanes', *Journal of Aeronautical Sciences* 3:4 (1936); G.E. Moore, 'Cramming more components onto integrated circuits', *Electronics* (19 April 1965); April 19, 1965; B. Nagy et al., 'Statistical basis for predicting technological progress', *PLoS ONE* 8(2): e52669.

26   B. Nagy et al., 'Statistical basis for predicting technological progress', PLOS 8:2 (February 2013), doi 10.1371/journal.pone.0052669; E. Rubin et al., 'A review of learning rates for electricity supply technologies', *Energy Policy* 86 (November 2015): 198–218.

27   N.M. Haegel et al., 'Terawatt-scale photovoltaics: Trajectories and challenges', *Science* 14:356 (April 2017): 141–43.

28   International Renewable Energy Agency, 'Renewable energy now accounts for a third of global power capacity', 2 April 2019, www.irena.org

29   Transport and Environment, 'EVs will be cheaper than petrol cars in all segments by 2027, BNEF analysis finds', 10 May 2021, www.transportenvironment.org

30   See Climate Town, 'It's time to break up with our gas stoves', 19 November 2021, https://www.youtube.com/watch?v=hX2aZUav-54

31   B. Seals and A. Krasner, *Gas Stoves: Health and Air Quality Impacts and Solutions*, RMI, 2020, www.rmi.org

32   H. Bambrick et al., 'Kicking the gas habit: how gas is harming our health', Climate Council, 6 May 2021, www.climatecouncil.org.au

33   World Steel Association, *Fact sheet: energy use in the steel industry*, www.worldsteel.org

34   G. Bhutanda, 'All the metals we mine each year, in one visualization', World
     Economic Forum, 14 October 2021, www.weforum.org
35   R. Campbell, E. Littleton and A. Armistead, 'Fossil fuel subsidies in Australia',
     Australia Institute, April 2021, australiainstitute.org.au
36   See fossilfuelsubsidytracker.org
37   J.L. Martire, 'Powering Indigenous communities with renewables', Renew,
     20 April 2020, renew.org.au
38   D. Carroll, 'State-owned utility reveals desire to disconnect rural town from
     grid', *PV Magazine*, 19 October 2021.
39   T. Moore et al., 'Better building standards are good for the climate, your
     health and your wallet. Here's what the National Construction Code
     could do better', *The Conversation*, 4 October 2021,
     www.theconversation.com.
40   Federal Reserve, *Report on the Economic Well-Being of US Households in 2019*,
     May 2020, www.federalreserve.gov
41   Energy Information Administration, 'One in three US households faces a
     challenge in meeting energy needs', 19 September 2018, www.eia.gov
42   K. Pabortsava and R.S. Lampitt, 'High concentrations of plastic hidden
     beneath the surface of the Atlantic Ocean', *Nature Communications* 11 (2020),
     doi 10.1038/s41467-020-17932-9; D. Pizzol et al., 'Pollutants and sperm
     quality: a systematic review and meta-analysis', *Environmental Science and
     Pollution Research* 28:4 (January 2021), doi 10.1007/s11356-020-11589-z; B.E.
     Lapointe et al., 'Nutrient content and stoichiometry of pelagic sargassum
     reflects increasing nitrogen availability in the Atlantic basin', *Nature
     Communications* 12 (2021), doi 10.1038/s41467-021-23135-7; C. Möllmann et al.,
     'Tipping point realized in cod fishery', *Scientific Reports* 11 (2021), doi 10.1038/
     s41598-021-93843-z; A. Torres et al., 'Assessing large-scale wildlife responses to
     human infrastructure development', *PNAS* 113:30 (26 July 2016), doi 10.1073/
     pnas.1522488113.
43   Y.M. Bar-On et al., 'The biomass distribution on Earth', *PNAS* 115(25): 6506–11.

Dr Saul Griffith is an inventor, entrepreneur, engineer, and author of *Electrify: An Optimist's Roadmap to Our Clean Energy Future* (MIT Press). He is the co-founder and chief scientist at Rewiring Australia and Rewiring America, nonprofits dedicated to decarbonising those countries (and the world) by electrifying everything. He is also founder and chief scientist at Otherlab, a research and development venture studio that has incubated numerous technology companies in sustainable energy and robotics. He was a recipient of a MacArthur 'Genius Grant' in 2007. He completed his PhD at Massachusetts Institute of Technology in 2004.

Saul lives in Wollongong, Australia, with his wife and two children and maintains his company offices in San Francisco.